DRUG COMPOUNDING PHARMACIES

RISKS AND OVERSIGHT ISSUES

Pharmacology - Research, Safety Testing and Regulation

Additional books in this series can be found on Nova's website under the Series tab.

Additional E-books in this series can be found on Nova's website under the E-book tab.

Public Health in the 21st Century

Additional books in this series can be found on Nova's website under the Series tab.

Additional E-books in this series can be found on Nova's website under the E-book tab.

PHARMACOLOGY - RESEARCH, SAFETY TESTING AND REGULATION

DRUG COMPOUNDING PHARMACIES

RISKS AND OVERSIGHT ISSUES

SETH SHARPE
EDITOR

nova publishers
New York

For permission to use material from this book please contact us:
Telephone 631-231-7269; Fax 631-231-8175
Web Site: http://www.novapublishers.com

Library of Congress Cataloging-in-Publication Data

ISBN: 978-1-62808-176-3

Published by Nova Science Publishers, Inc. † New York

CONTENTS

PREFACE

In light of the 2012 fungal meningitis outbreak, believed to have been caused by a contaminated compounded steroid injection, the regulation of human drug compounding has received significant attention. Drug compounding in its traditional form is the process of combining, mixing, or altering ingredients in order to create a medication for a particular patient. However, as illustrated by the entity that created the steroid medication linked with the meningitis outbreak, concerns have been raised about compounding pharmacies producing drugs on a larger scale. While drug compounding has historically been the focus of state governments through their regulation of pharmacies, questions have arisen regarding the extent the federal government can regulate the practice of compounding through the Food, Drug, and Cosmetic Act (FDCA). This book examines the FDA's regulation of drug compounding and discusses relevant legal authorities, and the potential limits to the FDA's authority to regulate human drug compounding.

Chapter 1 – Federal authority over compounding largely stems from the FDCA, enacted in 1938, and its subsequent amendments, including the Food and Drug Administration Modernization Act (FDAMA) of 1997, which added compounding specific provisions to the FDCA in Section 503A. A First Amendment challenge to Section 503A's advertising provisions and a resulting Supreme Court decision has created a split amongst the federal courts regarding whether Section 503A has any legal effect. The legal uncertainty regarding Section 503A, coupled with Section 503A's basis in the FDCA's "new drug," "adulteration," and "misbranding" provisions lends to an examination of the scope of federal authority under the base statute. Courts appear to agree that the federal government can regulate compounding activity that is akin to manufacturing, and courts have afforded deference to the FDA's interpretation of when a compounder is acting like a manufacturer. However, uncertainty remains regarding the possible limits to the FDA's power to regulate traditional compounding activities. This report will examine the FDA's regulation of drug compounding and will discuss relevant legal authorities. The report will conclude by discussing potential limits to the FDA's authority to regulate human drug compounding.

Chapter 2 – In light of the recent meningitis outbreak, believed to have been caused by a contaminated compounded steroid injection, the regulation of drug compounding has received significant attention. Compounding is traditionally defined as a process of combining, mixing, or altering ingredients in order to create a medication for a particular patient. However, as in the case of the pharmacy that produced the steroid medication, concerns have been raised about compounding pharmacies producing drugs on a larger scale, something more akin to drug manufacturing. While drug compounding has historically been regulated primarily by states through their regulation of pharmacies, recent questions have been raised about the extent to which the Federal Food, Drug, and Cosmetic Act (FFDCA) governs this practice, and what authority the U.S. Food and Drug Administration (FDA) has to regulate a compounded drug as a "new drug," subject to approval by the FDA, as well as other requirements.

In 1997, Congress enacted the FDA Modernization Act of 1997 (FDAMA), which was a comprehensive revision of the FFDCA. Section 127 of FDAMA added Section 503A to the FFDCA, which excepted compounded drugs from various "new drug" requirements, conditioned upon the compounded drugs meeting a variety of restrictions. One of the restrictions in Section 503A of the FFDCA was that drug providers were prohibited from soliciting or advertising particular compounded drugs. These speech restrictions were challenged on First Amendment grounds and were struck down by the Supreme Court in Thompson v. Western States Medical Center. Following this decision, there was controversy over the current status of compounded drugs under the FFDCA and whether the remaining provisions of Section 503A remain good law, an issue that the Supreme Court did not address in Western States. The two circuits that addressed this issue took different positions. While the Ninth Circuit in Western States determined that Section 503A was struck down in its entirety, the Fifth Circuit in Medical Center Pharmacy v. Mukasey found that the lawful provisions of Section 503A are still in effect. Accordingly, these cases have created an interesting scenario of non-uniform enforcement throughout the U.S. In the Fifth Circuit, compounded drugs are specifically exempted from new-drug, adulteration, and misbranding requirements of the FFDCA if certain criteria are met; while in the Ninth Circuit (and, according to the FDA, the rest of the United States), compounded drugs are subject to these requirements, but the FDA may exercise discretion in taking action against an entity that violates these provisions. This report provides a brief historical overview of the FDA's regulation of drug compounding and addresses these conflicting decisions. The report will also address the FDA's current authority to regulate compounded drugs under the FFDCA in light of these decisions.

Chapter 3 – An outbreak of fungal meningitis that has thus far claimed the lives of 53 people and sickened 733 brought national attention to compounding pharmacies, the drugs they produce, and the potential risks from these drugs. At the center of this outbreak was the Massachusetts-based New England Compounding Center (NECC) that produced the preservative-free injectable steroid drug that caused the deadly outbreak. This drug, found later to be contaminated with mold, was shipped to 23 states and injected into more than 14,000 patients, resulting in the worst pharmaceutical-related public health crisis in U.S. history. The NECC tragedy was not the first incident of compounding pharmacies causing fatalities or illnesses. Over the last decade, compounding pharmacies have been responsible for at least 23 additional deaths and 86 serious illnesses across the nation.

Unlike drug manufacturers, which are subjected to rigorous Food and Drug Administration (FDA) oversight, there is no requirement for compounding pharmacies to register with the FDA or to follow any FDA-specified procedures to ensure that the drugs they produce are safe and effective. In fact, the oversight of all pharmacies typically falls under the purview of individual states. Attempts by the FDA and by Congress to address safety concerns by ensuring uniform federal standards for compounding pharmacies have led to numerous lawsuits and conflicting judicial rulings and according to the FDA, created "ambiguity associated with FDA's authority over compounding pharmacies." The agency has called on Congress to take action to prevent another such outbreak, stating that if FDA's authority remains unchanged, "this type of incident will happen again. It's a matter of when, not if."

Chapter 4 – The recent outbreak of fungal meningitis caused by contaminated injectable steroids manufactured by New England Compounding Center (NECC) has currently killed 25 people and sickened 344 people in 18 states. It has also raised many questions, starting with a fundamental one: How could this have happened? This report describes the nature of the regulatory oversight and gaps in legal authority that led to one of the worst public health tragedies in recent U.S. history.

Traditional pharmacy compounding occurs when a pharmacist "combines, mixes, or alters various drug ingredients to create a medication for an individual patient in response to a practitioner's prescription."[1] The practice has typically been used to provide a patient with a medication that is not commercially available, such as adding flavor to a child's dose or removing an allergen from a drug for a patient with an allergy.

Chapter 5 – On September 26, 2012, as a result of the rapid work of the Tennessee Department of Public Health and the Centers for Disease Control and Prevention (CDCP), an outbreak of an unusual strain of fungal meningitis was identified.

Preservative free methylprednisolone acetate (MPA), administered via spinal injection, was quickly identified as a likely source of the infections. The MPA was traced back to a compounding pharmacy in Framingham Massachusetts, the New England Compounding Pharmacy Inc., doing business as the New England Compounding Center (NECC). The Food and Drug Administration (FDA) subsequently determined that three separate lots of MPA, totaling over 17,000 doses produced by NECC between May 21, 2012 and August 10, 2012, were contaminated with the *exserohilum rostratum* fungus.

To date, NECC's failure to produce a sterile and safe product has led to more than 30 deaths and over 450 serious illnesses requiring treatment with high risk anti-fungal medications. The efforts of the CDCP and the Tennessee Department of Public Health allowed public health officials in 23 states to rapidly track and begin monitoring the approximately 14,000 possible recipients of the contaminated drug. But thousands of people around the country continue to wait and see whether they will develop meningitis, joint infections, spinal abscesses, or arachnoiditis. Those treated will face the risk of kidney and liver damage from the powerful anti-fungal drugs.

While the quick work of the public health community has led to early identification and treatment of many cases of meningitis, and reduced the fatalities resulting from the administration of the contaminated MPA, the Committee's investigation demonstrates that this crisis should have, and could have, been avoided entirely.

Since its creation in 1998, inspections of NECC by state, federal, and independent investigators have identified and documented profound deficiencies in the company's production of sterile drugs. The company has also been cited on multiple occasions for improper use of prescription blanks to solicit orders and failure to comply with state regulations requiring patient-specific prescriptions for compounded drugs.

Moreover, the same drug at issue in the current outbreak, NECC-produced MPA, had previously been a suspected cause of at least two cases with bacterial meningitis-like symptoms. These reports triggered an FDA inspection of the facility ten years prior to the current outbreak, in August 2002.

While the FDA sampling of NECC-produced MPA proved sterile at the time, other MPA samples were found to contain bacteria. As an FDA employee stated in a power point presentation to the Massachusetts Board of Registration in Pharmacy (Board) at the time, "Sterilization techniques and aseptic practices continue to raise questions, despite no positive (nonsterile) results from latest samples. Absence of evidence is not evidence of absence."

Four years later in 2006, an independent evaluator reported to NECC manager and co-owner Barry Cadden that major areas of concern included "inadequate and incomplete documentation," that "end product testing is often performed on 'stock solutions' and not the end product that is required," "process controls including validation of sterilization cycles and media fills are inadequate," and "in many cases the procedures are not in strict accordance with USP 795/797" as required by Massachusetts state law.

In view of these repeated concerns with regard to the ability of NECC to safely produce compound drugs, it is difficult to understand why definitive action was never taken to either revoke its license or, at a minimum, closely monitor the company's operations. Instead, the company was allowed to grow and expand operations, ultimately holding licenses to ship drugs to at least 45 states. The same owners were subsequently permitted to open the far larger Ameridose, which supplied compounded drugs to hospitals around the country. Also, that company now has been found to lack adequate procedures to ensure that the compounds produced are safe, uniform or sterile.

This report is based on information obtained in the course of the Committee's investigation. It is intended to recount the known history of NECC, its related companies and their interactions with federal and state regulators as of November 15, 2012 to better understand the events leading to the current public health crisis.

Chapter 6 – In September 2012, the Centers for Disease Control and Prevention (CDC), the Tennessee Department of Health, the Food and Drug Administration (FDA), and other state health departments began investigating a rare, non-contagious outbreak of fungal meningitis. As of March 4, 2013, 20 states have reported 720 infections (including fungal meningitis and other conditions) and 48 deaths. Patients at risk are those who received injections of contaminated, preservative-free methylprednisone acetate produced by the

New England Compounding Center (NECC) after May 2012. On October 4, 2012, the FDA verified that NECC was the source of the contaminated products.

There has been extensive news coverage of the outbreak, which raised issues about the safety of compounded drugs and the role of federal and state governments in regulating compounded drugs and compounding pharmacies. A number of policy questions about FDA authority and resources were raised in congressional hearings held by the Energy and Commerce Committee of the U.S. House of Representatives and the Health, Education, Labor, and Pensions Committee of the U.S. Senate.

This report provides selected resources on the federal government's oversight of compounded drugs and compounding pharmacies, with an emphasis on relevant federal law and regulation and the activities of federal agencies related to and following the 2012 fungal meningitis outbreak. Resources from three states that have direct roles in this outbreak through licensing NECC, or reporting and tracking the fungal meningitis outbreak, are included as well. After a brief background section, the selected resources presented in this report appear in the following order:

- federal law;
- congressional hearings;
- proposed legislation (current and previous Congresses);
- congressional support agency products;
- FDA agency guidance and regulations, meetings, and related documents;
- FDA actions on drug compounding;
- other federal agency actions and resources;
- selected state resources; and
- articles from legal, professional, and scientific journals.

In: Drug Compounding Pharmacies
Editor: Seth Sharpe

ISBN: 978-1-62808-176-3

Chapter 1

FEDERAL AUTHORITY TO REGULATE THE COMPOUNDING OF HUMAN DRUGS*

Andrew Nolan

SUMMARY

In light of the 2012 fungal meningitis outbreak, believed to have been caused by a contaminated compounded steroid injection, the regulation of human drug compounding has received significant attention. Drug compounding in its traditional form is the process of combining, mixing, or altering ingredients in order to create a medication for a particular patient. However, as illustrated by the entity that created the steroid medication linked with the meningitis outbreak, concerns have been raised about compounding pharmacies producing drugs on a larger scale. While drug compounding has historically been the focus of state governments through their regulation of pharmacies, questions have arisen regarding the extent the federal government can regulate the practice of compounding through the Food, Drug, and Cosmetic Act (FDCA).

Federal authority over compounding largely stems from the FDCA, enacted in 1938, and its subsequent amendments, including the Food and Drug Administration Modernization Act (FDAMA) of 1997, which added compounding specific provisions to the FDCA in Section 503A. A First

* This is an edited, reformatted and augmented version of a Congressional Research Service publication, CRS Report for Congress R43038, prepared for Members and Committees of Congress, from www.crs.gov, dated April 12, 2013.

Amendment challenge to Section 503A's advertising provisions and a resulting Supreme Court decision has created a split amongst the federal courts regarding whether Section 503A has any legal effect. The legal uncertainty regarding Section 503A, coupled with Section 503A's basis in the FDCA's "new drug," "adulteration," and "misbranding" provisions lends to an examination of the scope of federal authority under the base statute. Courts appear to agree that the federal government can regulate compounding activity that is akin to manufacturing, and courts have afforded deference to the FDA's interpretation of when a compounder is acting like a manufacturer. However, uncertainty remains regarding the possible limits to the FDA's power to regulate traditional compounding activities. This report will examine the FDA's regulation of drug compounding and will discuss relevant legal authorities. The report will conclude by discussing potential limits to the FDA's authority to regulate human drug compounding.

BACKGROUND

Drug compounding traditionally refers to the process in which a pharmacist combines, mixes, or alters various ingredients to create a medication that is "tailored to the needs" of an individual patient.[1] Compounding is generally used to prepare medications that are not commercially available, such as a drug in a lower dosage for a child, or a drug without a dye or a preservative in response to a patient allergy.[2]

Recent events have refocused attentions on compounding and the federal role in regulating drug compounding. In the fall of 2012, the Food and Drug Administration (FDA) and the Centers for Disease Control and Prevention (CDC) identified the presence of bacterial and/or fungal contamination in steroid injections from a New England compounding pharmacy.[3] The contamination has been linked with a multistate fungal meningitis outbreak that resulted in over 50 deaths.[4]

As a consequence, questions have arisen regarding the safety of compounded drugs and the role of federal and state governments in regulating compounded drugs and compounding pharmacies. Congressional hearings held by the Energy and Commerce Committee of the U.S. House of Representatives and the Health, Education, Labor, and Pensions Committee of the U.S. Senate probed the current policy and legal questions prompted by the meningitis outbreak.[5] One of the issues raised at these hearings was the extent of federal authority over compounding.[6] During these hearings, the Commissioner of the FDA noted the need for clarification and strengthening of the FDA's authority to "prevent" contaminations caused by compounding from occurring in the future,[7] while other representatives of professional associations contended that existing FDA authorities were adequate to have prevented the 2012 outbreak.[8] Thus, in the wake of the 2012 meningitis outbreak, a central issue that this report explores is the nature and scope of federal authority to regulate entities that compound drugs for humans.

FEDERAL REGULATION OF COMPOUNDED DRUGS

Compounding Before the 1938 Food, Drug, and Cosmetic Act

At the inception of the modern federal drug law, compounding was practiced extensively and widely regulated by the states.[9] On June 30, 1906, Congress enacted the Pure Food and Drug Act,[10] the first significant drug legislation that attempted to address the problems of drug safety and effectiveness. However, the Pure Food and Drug Act, which was passed with the purpose to "secure the purity of food and drugs and to inform purchasers of what they are buying,"[11] did not mandate a federal premarket approval or notification system for new drugs, let alone attempt to substantively regulate drug compounding. Instead, the 1906 act was concerned with false labeling of food and drugs, and as long as a drug compounder made no false or misleading claims with regard to the ingredients in a product, the compounder's actions would not run afoul of the Pure Food and Drug Act.[12] The advent of modern federal regulation of drugs arrived 22 years later when Congress enacted the Food, Drug, and Cosmetic Act (FDCA).[13]

The 1938 Food, Drug, and Cosmetic Act

The Food, Drug, and Cosmetic Act of 1938 established a complex and comprehensive scheme for regulating drugs that are distributed in interstate commerce.[14] While the 1938 act on its face is silent with respect to compounding,[15] three provisions are particularly relevant to the subject. First, the act defines a "new drug" as "[a]ny drug ... not generally recognized ... as safe ... for use under the conditions prescribed, recommended, or suggested in the labeling thereof."[16] The statute forbids introducing any "new drug" into interstate commerce without FDA approval.[17] Under current law, to obtain FDA approval, the sponsor must submit an extensive application to the FDA containing data that demonstrate that the new drug is both safe and effective for its intended uses in humans and that it will be manufactured such that the drug will preserve its identity, strength, quality, and purity.[18]

In addition to the act's "new drug" provisions, the current version of the FDCA has "adulteration" provisions that deem a drug to be "adulterated" if, among other things, the drug and the methods used in its creation do not conform with "current good manufacturing practices" that ensure the product's safety and quality.[19] The FDA has promulgated regulations that describe current good manufacturing practices.[20] Those regulations include, for example, requirements that the equipment used in the manufacturing of drugs be cleaned or sterilized at certain times[21] and that records related to manufacturing practices be maintained for at least a year after the expiration date of a batch of drugs.[22]

Third, the 1938 act contains provisions that discuss when a drug is "misbranded."[23] In relevant part, a drug can be "misbranded" if the drug's labeling[24] fails to bear "adequate directions for use."[25] FDA regulations generally define "adequate directions for use" as directions under which a layperson can use a drug "safely and for which it is intended."[26] Federal regulations further note several examples of when a direction may be "inadequate," such as the omission of "all" purposes for which the drug is intended[27] or the incorrect specification of the time of administration of a drug.[28]

Section 301 of the FDCA prohibits an individual from introducing or delivering into interstate commerce (1) a "new drug" without prior FDA approval,[29] (2) an "adulterated" drug,[30] or (3) a "misbranded" drug.[31] That same section also prohibits the "adulteration" or "misbranding" of a drug in interstate commerce,[32] the "receipt in interstate commerce of" an adulterated or misbranded drug, and the manufacturing of any adulterated or misbranded drugs.[33] The FDCA authorizes the United States to seek injunctive relief in federal court to restrain any violations of Section 301,[34] and the law subjects individuals who have violated the FDCA's prohibitions to civil fines and criminal penalties.[35] Furthermore, Section 304 of the FDCA allows the United States to proceed against and seize mislabeled or adulterated drugs.[36]

The breadth of the FDCA's "new drug," "adulteration," and "misbranding" provisions, on their face, have the potential to touch on local, traditional compounding efforts.[37] However, for nearly 50 years following the passage of the FDCA, the FDA did not use the FDCA's "new drug," "adulteration," or "misbranding" provisions as a means to regulate the practice of compounding.[38] Instead, the "FDA generally left regulation of compounding to the States," and "[p]harmacists continued to provide patients with compounded drugs without applying for FDA approval of those drugs."[39] The rationale for the FDA's initial approach toward drug compounding was that it is widely recognized that compounded drugs could not meet the FDCA's "new drug" requirements.[40] As the Supreme Court has noted, subjecting compounding pharmacies to the statute's approval requirements "would as a practical matter, eliminate the practice of compounding, and thereby eliminate availability of compounded drugs for those patients who have no alternative treatment."[41] This is because FDA approval requires "proof of the safety and effectiveness of a new drug ... [as] established by rigorous, scientifically valid clinical studies," a process that is extremely time consuming and costly.[42] As the Court observed, it is neither medically practicable nor "economically feasible" to subject compounded medications to the type of testing required by the FDCA for a new drug approval;[43] "requiring such testing would force pharmacists to stop providing compounding drugs."[44]

Kefauver-Harris Drug Amendments of 1962

The Kefauver-Harris Drug Amendments of 1962 (Drug Amendments), enacted in the wake of the thalidomide crisis of the early 1960s, constituted the first mentioning of compounding in federal drug law.[45] In relevant part, the Drug Amendments required manufacturers and others to register their establishments and strengthened the ability of the FDA to conduct on-site inspections.[46] The Drug Amendments, however, exempted "pharmacies ... which do not ... compound, or process drugs for sale other than in the regular course of their business of dispensing or selling drugs at retail" from registration requirements[47] and from enhanced inspections of a pharmacy's books and records.[48] Case law interpreting the Drug Amendments of 1962, however, has held that federal authorities have the right to enter establishments involved in the production of drugs to inspect equipment, materials, containers, and labeling used in the production process.[49]

Food and Drug Administration Modernization Act of 1997 and Compounding

In 1997, Congress amended the FDCA through the Food and Drug Administration Modernization Act (FDAMA).[50] The amendment contained provisions on drug compounding that resulted, in part, from a growing concern that an "increasing number of establishments with retail pharmacy licenses [were] engaged in manufacturing and distributing unapproved new drugs for human use in a manner that was clearly outside the bounds of traditional pharmacy practice."[51]

Put another way, the FDA was concerned that "some pharmacists were manufacturing and selling drugs under the guise of compounding," as a way of avoiding the FDCA's "new drug," "adulteration," and "misbranding" provisions.[52] Accordingly, the intent of the FDAMA's compounding provisions was "to clarify the application of the [FDCA] to professional practice of pharmacists' compounding of drug products."[53]

To do so, the FDAMA explicitly deems that three provisions of the FDCA are inapplicable to drug compounding that complies with various requirements.[54]

The three provisions are the FDCA's provisions on (1) the sale of a "new drug," (2) adulteration and the need to adhere to "good manufacturing practices," and (3) misbranding and the need to provide "adequate directions for use" on a drug.[55] With respect to requirements for compounding, the law requires that compounding be done by a licensed pharmacist or physician in response to a valid prescription for an identified individual patient, or, if prepared before the receipt of such a prescription, the drug be made in only "limited quantities" and in response to a history of the licensed pharmacist's or physician's receipt of valid prescription orders for that drug product within an established relationship between the pharmacist, the patient, and the prescriber.[56]

Additionally, the law requires that the compounded drug be made from approved ingredients that meet certain standards, and the compounded drug may not appear on an FDA list of drug products that have been withdrawn or removed from the market because they were found to be unsafe or ineffective.[57] The FDAMA also prohibits the compounding of drugs in "inordinate amounts" such that the drugs are "essentially copies of a commercially available drug product."[58] The law also allows the FDA to identify certain drug products that present "demonstrable difficulties" for compounding in terms of safety or effectiveness, prohibiting the compounding of such articles.[59] In states that have not entered into an agreement with the FDA addressing the distribution of "inordinate amounts" of compounded drugs in interstate commerce, the pharmacy, pharmacist, or physician compounding the drug may not distribute compounded drugs out of state in quantities exceeding 5% of that entity's total prescription orders.[60]

FDAMA and Federal Authority after *Thompson v. Western States Medical Center*

The FDAMA also included a ban on advertising by compounders.[61] However, in 2003, the Supreme Court in *Thompson v. Western States Medical Center* invalidated the FDAMA's advertising provisions on First Amendment grounds.[62] Generally, if a part of a law is held to be unconstitutional, the unconstitutional provision is "severable" or "dropped" from what remains of a law unless it is "evident that the Legislature would not have enacted" the remaining provisions independently of the unconstitutional provision.[63] In the wake of *Western States,* a split has emerged amongst the federal circuit court of appeals about whether the remaining provisions of the FDAMA are "severable" from the advertising provision at issue in the *Western States* case.[64]

As a consequence, within the states within the Ninth Circuit,[65] the controlling precedent renders the remaining provisions of the FDAMA inoperable.[66] The opposite is true in the Fifth Circuit,[67] and the legal effect of the FDAMA is unresolved in the rest of the country.[68] The FDA noted in 2002, prior to the Fifth Circuit's decision, that in light of the Supreme Court decision, the FDAMA "is now invalid,"[69] and instead, the FDA "outlined the criteria"[70] it would use to assess "what types of compounding might be subject to enforcement under current law" (i.e., the FDCA)[71] in a 2002 Compliance Policy Guide (CPG).[72] In the wake of the Fifth Circuit's decision, the FDA has taken the position[73] that the agency "will apply the non-advertising provisions of section 503A to entities covered by this provision that are located within the jurisdiction of the Fifth Circuit ... as well as to the plaintiffs that brought the *Medical Center* case."[74]

The 2002 CPG was issued out of a "need for immediate guidance" for "the compounding industry on what factors the [FDA] will consider in exercising its enforcement discretion regarding pharmacy compounding."[75] The policy guidance generally notes while the FDA "will continue to defer to state authorities regarding less significant violations of the [FDCA] related to pharmacy compounding of human drugs,"[76] the agency will focus on "concerns normally associated with a drug manufacturer and result in significant violations of the new drug, adulteration, or misbranding provisions of the Act."[77] In determining whether to take an enforcement action against a compounding entity, the CPG notes nine non-exhaustive factors that will guide regulators.[78] The factors are summarized in *Table 1*.[79]

Table 1. Factors Guiding Whether the FDA Will Take Action Against a Compounder of Human Drugs

Is the compounding ... ?	If yes, is the FDA more likely to assert its regulatory authority?
... done in anticipation of receiving prescriptions?	No
... involving a drug that was withdrawn or removed from the market for safety reasons?[a]	Yes
... based from bulk active ingredients?	Yes (if not from a FDA-approved drug or without a FDA sanctioned New Drug Application)
... done without obtaining written assurance from the supplier that the drug substances were made in an FDA-registered facility?	Yes
... not in compliance with official compendia requirements?	Yes
... using commercial scale manufacturing or testing equipment?	Yes
... done for third parties who will resell the drugs?	Yes
... of drug products that are commercially available in the	Yes
marketplace or essentially copies of commercially available drugs?	
... failing to comply with applicable state law?	Yes

Source: Prepared by CRS.

Notes: Derived from FDA Compliance Policy Guide §460.200 (May 29, 2002).

[a] See 21 C.F.R. §216.24 (list of drug products withdrawn or removed from the market for reasons of safety or effectiveness).

In 2003, the FDA issued guidance with regard to veterinary compounding.[80] Similar to its 2002 guidance on human compounding, the FDA noted its concerns regarding compounding that is "clearly outside of the bounds of traditional pharmacy practice" and is "intended to circumvent [the FDCA's] drug approval process and provide for the mass marketing of products that have been produced with little to no quality control or manufacturing standards to ensure the purity, potency, and stability of the product."[81] The 2003 CPG announced that while the FDA will "generally" "defer to state authorities regarding the day-to-day regulation of compounding by veterinarians," federal authority will be asserted when the activities in question raise the "kinds of concerns normally associated with a drug manufacturer" and "result in significant violations of the new animal drug, adulteration, or misbranding provisions" of the act.[82] In determining whether to take an enforcement action against a compounding entity, the CPG on animal drug compounding notes 13 non-exhaustive factors that will guide regulators.[83] The factors largely mirror the concerns illustrated in the 2002 CPG for human drug compounding and include concerns common to both human and animal drug compounding.[84] For example, the FDA noted that the volume of drugs being compounded or whether the compounding used commercial scale manufacturing equipment would be used by federal officials when determining whether to take an enforcement action against an animal drug compounder.[85]

Both the 2002 and 2003 CPGs on drug compounding make clear that the FDA, as a matter of public policy, will focus its enforcement efforts on compounding efforts that are akin to "manufacturing" and will defer to state governments with respect to "traditional compounding."[86] As a matter of law, the FDA does not attempt, in the two CPGs, to limit its authority under the FDCA. Instead, the CPGs note that the FDCA does not "distinguish compounding from manufacturing"[87] and deem violations of the FDCA that do not relate to manufacturing to be a concern of "state authorities."[88]

LIMITS ON FEDERAL AUTHORITY TO REGULATE COMPOUNDED DRUGS

The legal uncertainty about the validity of the compounding provisions of the FDAMA, coupled with the FDA's renewed reliance on the FDCA's new drug, adulteration, and misbranding provisions as the source of the federal authority to regulate drug compounding, has prompted considerable discussion with respect to the extent of the FDA's authority under the 1938 act. Moreover, even if the FDAMA's compounding provisions still have legal effect, the provisions are based on blanket exemptions for certain compounding activity from the FDCA's new drug, adulteration, and misbranding requirements, all of which are based on the assumption that broad federal authority to regulate drug compounding exists.[89] With this backdrop, a debate has arisen regarding the limits of federal authority under the FDCA with respect to the regulation of drug compounding.

The FDA maintains the position that pharmacy-compounded drugs are "new drugs" or "new animal drugs" within the meaning of the FDCA,[90] and, therefore, the entire practice of compounding is "under FDA scrutiny."[91] According to the agency, as a "matter of discretion" the FDA has "historically declined" to take "enforcement action against pharmacies engaged in traditional drug compounding."[92] Pharmacists, in response, have disputed the reach of the FDA's authority, finding "cold comfort in the FDA's promised self-restraint."[93] Instead, those who oppose the FDA's theory of its authority to regulate compounding contend that the practice of compounding was widespread at the time of the 1938 act, and Congress would not have intended to expand the FDA's regulatory authority to make compounding activities covered by the act's new drug, adulteration, and misbranding provisions,[94] regulations that, if enforced literally, would functionally prohibit traditional compounding.[95]

Courts that have examined the issue have all agreed that the FDA has *some* authority to regulate compounding.[96] Specifically, case law appears to find that the FDCA does provide the federal government with the authority to prohibit pharmacists from manufacturing under the guise of compounding.[97] According to one recent court decision, "[t]o the extent that a pharmacist's bulk compounding activity moves beyond the bound of traditional compounding and begins to approximate the 'manufacturing' of unapproved drugs, there seems little question that this activity is squarely within the crosshairs of the FDCA."[98] Moreover, courts have interpreted the FDA's 2002 CPG as a reasonable means by which the FDA can gauge whether an entity is engaged in manufacturing.[99]

However, a question remains as to whether a so-called "maximalist" interpretation of the FDA's authority exists, such that any "traditional compounding" is "*per se* unlawful under the FDCA."[100] In favor of the maximalist interpretation of the FDA's authority over compounding is the plain language of the FDCA.[101] Specifically, the FDCA's "new drug" provisions broadly state that no person can "introduce or deliver for introduction into interstate commerce any new drug, unless an approval of an application filed" pursuant to the statute with the FDA has occurred with respect to the drug.[102] Moreover, Section 501 of the FDCA broadly deems a drug to be "adulterated" if it "is a drug and the" way it was manufactured, processed, packaged, or held does not conform with good manufacturing practices that ensure its quality and purity.[103] The FDCA's misbranding provisions are equally broad, in that the statute deems a drug to be "misbranded" if its labeling does not bear "adequate directions for use."[104] In none of the three aforementioned provisions does the FDCA's plain language exempt traditional compounding activity from the act's reach. As one court has noted, "[c]ompounded drugs are, after all, drugs,"[105] and therefore, the plain language of the statute would seem to indicate that compounding activities are within the purview of the FDA's authority under the act.[106] Moreover, the maximalist interpretation is supported by the argument that because exemptions exist under the FDCA's provisions for other types of drugs, such "investigational drugs,"[107] one can presume from Congress's refusal to create a general exemption for traditional compounding an implicit extension of federal authority over all forms of compounding.[108]

On the other hand, given the "ubiquity of pharmacy compounding at the time Congress passed the FDCA," it can be argued that it would be absurd for Congress to have taken the "unprecedented" action to implicitly regulate compounded drugs through the 1938 act.[109] This argument is buttressed by Supreme Court case law on statutory interpretation that generally holds that it should not be assumed that Congress "alter[s] the fundamental details of a regulatory scheme in vague terms or ancillary provisions."[110] Moreover, the legislative history of the FDCA appears to support the view that manufacturers were the intended target of the regulatory scheme imposed by the 1938 act.[111] For example, the then-President of the American Pharmaceutical Association testified before a Senate subcommittee in 1938 in support of the legislation that while the "regulations governing ... the practice of pharmacy are very strict ... the privileges of unlicensed persons operating outside of pharmacies are so extensive that the public enjoys little protection,"[112] evincing that Congress's central concern with the FDCA was manufacturing, not traditional pharmacy compounding. Legislative history can guide a court in assessing the correct interpretation of the scope of the FDA's authority under the act.[113]

The current limits of federal authority to regulate traditional drug compounding have been discussed by very few courts, and each court that has approached the issue did so from a unique factual setting that colored the eventual outcome of the case.[114] As a consequence, the precise limit to federal authority with respect to drug compounding remains uncertain, although Congress could arguably expand the scope of the FDCA to reach traditional compounding.[115] As a matter of policy, the FDA has generally declined to test the current limits of its authority to regulate compounding,[116] preferring to defer to state governments with respect to the regulation of "traditional compounding."[117] As a result, absent further congressional action, the limits of the FDA's authority to regulate all forms of compounding will likely continue to be unresolved.[118] With respect to enforcement efforts clearly within the FDA's broad discretion, such as combating compounding that is akin to large-scale manufacturing, courts will likely provide considerable deference to the FDA.[119]

End Notes

[1] See Thompson v. W. States Med. Ctr., 535 U.S. 357, 360-61 (2002).

[2] Id. at 361.

[3] Centers for Disease Control and Prevention, Multistate Fungal Meningitis Outbreak Investigation—Current Case Count, http://www.cdc.gov/hai/outbreaks/meningitis.html.

[4] Centers for Disease Control and Prevention, Multistate Fungal Meningitis Outbreak Investigation—Current Case Count, http://www.cdc.gov/hai/outbreaks/meningitis-map-large.html (noting that as of March 25, 2013, 730 cases of fungal infections had been linked to steroid injections, resulting in 51 deaths).

[5] See U.S. Congress, House Committee on Energy and Commerce, Subcommittee on Oversight and Investigations, The Fungal Meningitis Outbreak: Could It Have Been Prevented, 112th Cong., 2nd sess., November 14, 2012; see also U.S. Congress, Senate Committee on Health, Education, Labor, and Pensions, Pharmacy Compounding: Implications of the 2012 Meningitis Outbreak, 112th Cong., 2nd sess., November 15, 2012, http://www.help.senate.gov/hearings/hearing/?id=5f5def0d-5056-a032-5297-eab57634d 209.

[6] U.S. Congress, Senate Committee on Health, Education, Labor, and Pensions, Pharmacy Compounding: Implications of the 2012 Meningitis Outbreak, 112th Cong., 2nd sess., November 15, 2012, http://www.help.senate.gov/hearings/ hearing/?id=5f5def0d-5056-a032-5297-eab57634d209.

[7] See Margaret A. Hamburg, Testimony before U.S. Congress, Senate Health, Education, Labor and Pensions Committee, Pharmacy Compounding: Implications of the 2012 Meningitis Outbreak, 113th Cong., November 15, 2012, http://www.help.senate.gov/imo/media/doc/Hamburg3.pdf ("As described above, FDA's ability to take action against compounding that exceeds the bounds of traditional pharmacy compounding and poses risks to patients has been hampered by gaps and ambiguities in the law, which have led to legal challenges to FDA's authority to inspect pharmacies and take appropriate enforcement actions.")

[8] See, e.g., David Miller, International Academy of Compounding Pharmacists, Testimony before U.S. Congress, Senate Health, Education, Labor and Pensions Committee, Pharmacy Compounding: Implications of the 2012 Meningitis Outbreak, 113th Cong., November 15, 2012, http://www.help.senate.gov/imo/media/doc/Miller8.pdf the federal standpoint, the FDCA's existing inspection provision, section 704, allows FDA oversight when a pharmacy is not operating in conformity with governing state laws, or akin to a drug manufacturer.")

[9] See Joint Session of Am. Pharm. Ass'n, et al., 17 J. Am. Pharm. Ass'n 1000, 1010-13 (1938).

[10] P.L. 59-384, 34 Stat. 768 (1906).

[11] United States v. Antikamnia Chem. Co., 231 U.S. 654, 665 (1913)

[12] See James Harvey Young, The Toadstool Millionaires: A Social History of Patent Medicines in America Before Federal Regulation 9, 243-44 (Princeton Univ. Press. 1961).

[13] P.L.75-717, 52 Stat. 1057 (1938).

[14] See generally CRS Report R41983, How FDA Approves Drugs and Regulates Their Safety and Effectiveness, by Susan Thaul.

[15] United States v. Franck's Lab, Inc., 816 F. Supp. 2d 1209, 1254 n.7 (M.D. Fl. 2011) (noting that the FDCA is "silent on the topic" of compounding).

[16] 21 U.S.C. §321(p)(1). The "new drug" provision of the FDCA contains a "grandfather clause" that excludes drugs subject to regulation under a 1906 federal statute, if the labeling is appropriate. Id. The FDCA broadly defines the term "drug" to include "articles intended for use in the diagnosis, cure, mitigation, treatment or prevention of disease in man or other animals." 21 U.S.C. §321(g)(1)(B).

[17] 21 U.S.C. §§331(d), 355(a).

[18] 21 U.S.C. §§355(a)-(d). The FDCA has specific provisions that discuss the role of new animal drugs. Specifically, "new animal drugs" that have not been approved by the FDA are "unsafe" and "adulterated." See 21 U.S.C. §§351(a)(5) & 360b(a)(1)(A).

[19] 21 U.S.C. §351(a)(1)(B). The precursor to the modern language regarding "good manufacturing practices" in the 1938 Act were provisions that broadly deemed a drug to be "adulterated" if it had been "prepared, packed, or held under insanitary conditions ..." See P.L.75-717, §501(a)(2). The provision on "good manufacturing practices" resulted from the Kefauver-Harris Drug Amendments of 1962. See infra.

[20] See 21 C.F.R. parts 210, 211, 225, and 226.

[21] See 21 C.F.R. §211.67.

[22] See 21 C.F.R. §211.180.

[23] See 21 U.S.C. §352.

[24] Labeling is defined under the FFDCA to include "all labels and other written, printed, or graphic matters (1) upon any article or any of its containers or wrappers, or (2) accompanying such article." 21 U.S.C. §321(m). The Supreme Court, in Kordel v. United States, broadly interpreted the phrase "accompanying such article" to include materials that supplement or explain the product. 335 U.S. 345, 349 (1948).

[25] 21 U.S.C. §352(f)(1). The statute does allow the Secretary to promulgate regulations exempting a drug if providing "adequate directions for use" are not necessary for the "protection of public health." Id. 21 C.F.R. §201.122 exempts from the FDCA's "adequate directions for use" requirement drugs in bulk form intended for processing, repackaging or use in the "manufacture of another drug" if the bulk drug contains labeling that indicates its proper use. Id. However, the exemption only extends to a drug where an approved new drug application or new animal drug application or new animal drug index listing is affiliated with the production. See 21 C.F.R. 201.122(a)-(c).

[26] 21 C.F.R. §201.5.

[27] 21 C.F.R. §201.5(a).

[28] 21 C.F.R. §201.5(e). For example, a drug label could be inadequate if it incorrectly specifies that the drug should be taken with food. Id.

[29] 21 U.S.C. §301(d).

[30] 21 U.S.C. §301(a).

[31] 21 U.S.C. §301(a).

[32] 21 U.S.C. §301(b).

[33] 21 U.S.C. §301(c).

[34] 21 U.S.C. §332(a).

[35] See 21 U.S.C. §333.

[36] 21 U.S.C. §334(a).

[37] See Medical Ctr. Pharm. v. Mukasey, 536 F.3d 383, 395 (5th Cir. 2008) (noting the language of the FDCA's new drug definition to be "both plain and expansive"); United States v. Franck's Lab, Inc., 816 F. Supp. 2d 1209, 1237 (M.D. Fl. 2011) (noting that the FDCA's provision if "read literally" would apply to bulk compounding).

[38] See W. States, 535 U.S. at 362.

[39] Id.

[40] Statement of Steven K. Galson, Acting Director, Center for Drug Evaluation and Research, U.S. Food and Drug Administration, before the Senate Committee on Health, Education, Labor and Pensions, Hearing on "Federal and State Role in Pharmacy Compounding and Reconstitution: Exploring the Right Mix to Protect Patients," (October 23, 2003). See also Western States, 535 U.S. at 369 ("...it would not make sense to require compounded drugs created to meet the unique needs of patients to undergo the testing required for the new drug approval process. Pharmacists do not make enough money from small-scale compounding to make safety and efficacy testing of their compounded drugs economically feasible, so requiring such testing would force pharmacists to stop providing compounded drugs.")

[41] W. States, 535 U.S. at 369.

[42] Id. at 368-69.

[43] 21 U.S.C. §355.

[44] W. States, 535 U.S. at 370.

[45] P.L. 87-781, 76 Stat. 780 (1962) (current version as amended at 21 U.S.C. §§301-392).

[46] Id.; see generally Bryan Christopher Moody, "Prescription Medication and Consumer Protection: A Time for Reform," The Journal of Pharmacy & Law, vol. 5 (1996), p. 21 ("Additionally, the Drug Amendments of 1962 required the following: that drug companies register with the FDA and be inspected by the FDA every two years, that all drug company records be made available for inspection, that adverse drug reactions be transmitted to the FDA, that the FDA be given full authority to regulate the advertising of prescription drugs and all advertising must relate to the adverse effects and contraindications to consumer, and that investigational drug studies require informed patient consent prior to trials on humans.")

[47] See 21 U.S.C. §360(g)(1).

[48] See 21 U.S.C. §374(a)(2)(A).

[49] See Wedgewood Vill. Pharm., Inc. v. United States, 421 F.3d 263, 269 (3d Cir. 2005) (holding that a compounder was not exempt from non-record inspections under 21 U.S.C. §374(a)).

[50] P.L. 105-115, 111 Stat. 2296 (1997). The FDAMA was a product of a 1992 FDA Compliance Policy Guide that outlined the agency's policies regarding drug compounding. See W. States., 535 U.S. at 362-64 ("Congress turned portions of this policy into law when it enacted the FDAMA in 1997).

[51] CPG §7132.16 (Mar. 1992).

[52] W. States, 535 U.S. at 362.

[53] S.Rept. 105-43, Committee on Labor and Human Resources, p. 67 (July 1, 1997).

[54] See 21 U.S.C. §353a(a).

[55] Id. (citing Sections 501(a)(2)(B) (adulteration), 502(f)(1) (misbranding), 505 (new drugs)).

[56] 21 U.S.C. §353a(a).

[57] 21 U.S.C. §§353a(b)(1)(A)-(C).

[58] 21 U.S.C. §353a(b)(1)(D). The definition of the term "inordinate amounts" is delegated in the statute to the Secretary of Health and Human Services. Id. FDA regulations have not yet formally defined the term "inordinate amounts." See 21 C.F.R. §216.23 et seq.

[59] 21 U.S.C. §353a(b)(3)(A).

[60] 21 U.S.C. §353a(b)(3)(B).

[61] 21 U.S.C. §353a(c).

[62] 535 U.S. at 360.

[63] See Alaska Airlines, Inc. v. Brock, 480 U.S. 678, 684 (1987) (internal quotations and citations omitted).

[64] Compare W. State Med Ctr. v. Shalala, 238 F.3d 1090, 1096-98 (9th Cir. 2001), aff'd on other grounds, 535 U.S. 357 (2002), with Medical Ctr. Pharmacy v. Mukasey, 536 F.3d 383, 401(5th Cir. 2008). The Supreme Court in Western States declined to rule on the issue of severability. 535 U.S. at 366 ("Because neither party petitioned for certiorari on the severability issue, we have no occasion to review that portion of the Court of Appeals decision.") For a more in depth discussion on these two cases and the resulting circuit split, see CRS Report R40503, FDA's Authority to Regulate Drug Compounding: A Legal Analysis, by Jennifer Staman.

[65] The Ninth Circuit Court of Appeals covers the states of Alaska, Arizona, California, Idaho, Montana, Nevada, Oregon, Washington, and Hawaii. See 28 U.S.C. §41.

[66] W. State Med Ctr., 238 F.3d at 1096-98.

[67] Medical Ctr. Pharmacy, 536 F.3d at 401. The Fifth Circuit's jurisdiction is limited to the states of Louisiana, Mississippi, and Texas. See 28 U.S.C. §41.

[68] An unpublished federal district court decision in Colorado, faced with both parties in agreement that the FDAMA's advertising provisions were severable from the rest of the act, opined that the Fifth Circuit's decision with respect to severability was not "so palpably incorrect that applying its rule would be clearly be error." See United States v. Bader, No. 07-CR-338, 2009 WL 2219258 at *9 (D. Colo. July 23, 2009), aff'd 678 F.3d 858 (11th Cir. 2012). On appeal to the Tenth Circuit, the appellate court found the Fifth Circuit's decision to be inapplicable to the case at hand as the case did not implicate any provisions of the FDAMA. 678 F.3d at 888-89 ("Consequently, Medical Center Pharmacy is inapposite.")

[69] Compliance Policy Guide §460.200 (May 29, 2002).

[70] Wedgewood Vill. Pharm., Inc. v. United States, 421 F.3d 263, 272 (3d Cir. 2005).

[71] Bader, 678 F.3d at 888 (noting that the "overarching premise of [the 2002 CPG on compounding] was that, generally, the FDA had maintained some degree of regulatory authority over pharmacy compounding both before Western States ... and the FDAMA").

[72] CPG §460.200 (May 29, 2002); see generally United States v. Livdahl, 459 F. Supp. 2d 1255, 1263 (S.D. Fla. 2005) (noting that "in the wake of Western States, the FDA issued a new [CPG] dealing with compounded drugs [similar to the one codified in the FDAMA] ... [t]his new policy is unaffected by the decision in Western States).

[73] See Food and Drug Administration, "Medical Center Pharmacy v. Mukasey," press release, May 6, 2009, http://www.fda.gov/Drugs/GuidanceComplianceRegulatoryInformation/PharmacyCompounding/ucm134919.htm ("In the meantime, FDA will follow the court's decision in the Fifth Circuit and with respect to the plaintiffs covered by the decision. Elsewhere, the agency will continue to follow the enforcement approach reflected in the Compliance Policy Guide (CPG) section 460.200 [Pharmacy Compounding] issued by FDA on May 29, 2002."); see also Letter from Alonza E. Cruse, Director, Los Angeles District, U.S. Food and Drug Administration, to Ari S. Schafer, Civic Center Pharmacy (December 16, 2008), available at http://www.fda.gov/ICECI/EnforcementActions/ WarningLetters/ 2008/ ucm1048048.htm; see also Letter from Alonza E. Cruse, Director, Los Angeles District, U.S. Food and Drug Administration, to Charles T. Bonner, Steven's Pharmacy (November 12, 2008), available at http://www.fda.gov/ ICECI/EnforcementActions/ WarningLetters/2008/ucm1048074.htm.

[74] The practical effect of whether the FDAMA's compounding provisions apply or whether the CPG applies is not readily apparent, as both the compounding provisions of the FDAMA and the CPG are rooted in authority provided by the FDCA. Compare 21 U.S.C. §503A(a) (stating that Sections 501(a)(2)(B), 502(f)(1) and 505 of the FDCA do not apply to certain

compounding activity) with CPG §460.200 (May 29, 2002) (providing guidance on "what types of compounding might be subject to enforcement action under current law").

[75] CPG §460.200 (May 29, 2002).

[76] Id.

[77] Id.

[78] Id.

[79] For a more in depth discussion on the 2002 FDA CPG on compounding, see CRS Report R40503, FDA's Authority to Regulate Drug Compounding: A Legal Analysis, by Jennifer Staman.

[80] See CPG §608.400 (July 14, 2003). The FDAMA's provision on compounding addressed human, not animal, use. See 21 U.S.C. §353a(a).

[81] See CPG §608.400 (July 14, 2003).

[82] Id. The "new animal drug" provisions of the FDCA are similar, but not identical, to the provisions for human drugs. Pursuant to Section 512 of the FDCA, absent FDA approval, a "new animal drug" is "deemed unsafe, which, in turn, renders the drug "adulterated" under Section 501(a) and "misbranded" under Section 502(f) unless the drug bears "adequate directions for use." See 21 U.S.C. §360b. Hence, similar "to human drugs, any drug ... the composition of which has not already been approved by the FDA constitutes a 'new animal drug' within the meaning of the statute.'" See Medical Ctr. Pharm., 536 F.3d at 407 (internal citations omitted).

[83] See CPG §460.200 (May 29, 2002); see also CPG §608.400 (July 14, 2003).

[84] Id.

[85] Id. While animal drug compounding is beyond the scope of this report, the overlap in the legal framework for human and animal drugs, see supra footnote 82, make the FDA's 2003 guidance useful to gauge the agency's general approach toward the regulation of drug compounding.

[86] Id.

[87] Id.

[88] See CPG §460.200 (May 29, 2002). The CPG, as a general policy statement, as opposed to a rule that affords the public notice and an opportunity to comment on the rule, does not have the force of law,88 and does not receive the fullest deference from courts of law. See United States v. Mead Corp., 533 U.S. 218, 229 (2001). Instead, a CPG "merely provide[s] guidance for FDA agents in determining possible violations, within their discretion." Professionals & Patients for Customized Care v. Shalala, 847 F. Supp. 1359, 1365 (S.D. Tex. 1994). For a lengthier discussion of the legal effect of the CPG, see CRS Report R40503, FDA's Authority to Regulate Drug Compounding: A Legal Analysis, by Jennifer Staman.

[89] 21 U.S.C. §353a(a) (exempting certain compounding activities from the reach of the FDCA Sections 501(a)(2)(B) (adulteration), 502(f)(1) (misbranding), 505 (new drugs)); but see Medical Ctr. Pharm., 536 F.3d at 405 (noting that because Congress started from a default premise that the FDCA's "adulteration, misbranding, and new drug provisions apply to – and thereby restrict – all drugs created by any means," including compounding, that the FDCA must reach compounding). However, it can be argued that the understanding of a future Congress regarding previously enacted legislation will "rarely override a reasonable interpretation of a statute that can be gleaned from its language." See Consumer Prod. Safety Comm'n v. GTE Sylvania, Inc., 447 U.S. 102, 118 n.13 (1980).

[90] Brief of the Appellant United States, United States v. Franck's Lab, Inc., 11-15350-BB (11th Cir 2012). at pg. 26.

[91] Food and Drug Administration, "Is It Really FDA Approved?," press release, February 20, 2009, http://www.fda.gov/downloads/ForConsumers/ConsumerUpdates/UCM143301.pdf.

[92] Brief of the Appellant United States, United States v. Franck's Lab, Inc., 11-15350-BB (11th Cir 2012). at pg. 26; see also Professionals & Patients for Customized Care v. Shalala, 56 F.3d 592, 593-594 (5th Cir. 1995).

[93] Medical Ctr. Pharm., 536 F.3d at 399.

[94] Id.

[95] See W. States Med. Ctr., 535 U.S. at 369 ("[R]equiring such testing would force pharmacists to stop providing compounded drugs.")

[96] See generally Franck's Lab, 816 F. Supp. 2d at 1235-1239 (discussing the cases that have examined the scope of the FDA's authority with respect to compounding).

[97] Id.; see also Med. Ctr. Pharm., 536 F.3d at 399 ("Construing the FDCA to give the FDA authority over compounding would thus not necessarily 'lead to a result so bizarre that Congress could not have intended it.'")(internal citations omitted); In re Wedgewood Vill. Pharmacy, 270 F. Supp. 2d 525, 549 (D.N.J. 2003) ("Congress intended that the FDCA, both in its original form and as amended, allow the FDA broad enforcement powers to fulfill its mandate that it protect the public from unsafe medication.")

[98] Franck's Lab, 816 F. Supp. 2d at 1246.

[99] See Wedgewood Vill. Pharm, 421 F.3d at 272 ("We agree that the factors set forth in the CPG are reasonable and that they reflect the FDA's 'careful consideration ... over a long period of time.") (internal citations omitted).

[100] See Franck's Lab, 816 F. Supp. 2d at 1239.

[101] The plain language of a statute is the starting point for statutory interpretation, and if a statute's terms are "plain," the Supreme Court has noted that a court should look no further and enforce the law "according to its terms." See Hartford Underwriters Ins. Co. v. Union Planters Bank, N.A., 530 U.S. 1, 6 (2000).

[102] 21 U.S.C. §355(a).

[103] 21 U.S.C. §351(a)(1)(B).

[104] 21 U.S.C. §352(f)(1).

[105] Medical Ctr. Pharm., 536 F.3d at 395.

[106] Id.; see also United States v. 9/1 Kg. Containers, 854 F.2d 173, 176 (7th Cir. 1988) (holding, with respect to the misbranding provisions of the FFDCA, "Congress gave the FDA comprehensive powers to license the manufacture of drugs and limit theirs sales," leaving it to the discretion of the FDA to regulate bulk compounding); United States v. Algon Chemical, Inc., 879 F.2d 1154, 1163 (3d Cir. 1989) (concluding that Congress did not intend that bulk drug compounding "be excluded from the Act's labeling requirements").

[107] 21 U.S.C. §360b(j).

[108] Medical Ctr. Pharm., 536 F.3d at 395; but see Franck's Lab, 816 F. Supp. 2d at 1244 ("However, if we were to presume a delegation of power from the absence of an express withholding of such power, agencies would enjoy virtually limitless hegemony") (internal citations and quotations omitted).

[109] Id. at 398 (summarizing the pharmacists' position); see also Franck's Lab, 816 F. Supp. 2d at 1243-44.

[110] See Whitman v. American Trucking Assoc., 531 U.S. 457, 458 (2001) ("[Congress] does not, one might say, hide elephants in mouseholes"); see also FDA v. Brown & Williamson Tobacco Corp., 529 U.S. 120, 160 ("[W]e are confident that Congress could not have intended to delegate a decision of such economic and political significance to an agency in so cryptic a fashion.")

[111] Franck's Lab, 816 F. Supp. 2d at 1245-46.

[112] Food, Drugs and Comsetis: Hearings Before a Cubcomm. of the Comm. on Commerce, 74 Cong. 100, 102 (1935) (statement of Robert P. Fischelis, president American Pharmaceutical Ass'n).

[113] See generally Crandon v. United States, 494 U.S. 152, 160 (1990).

[114] See generally Franck's Lab, 816 F. Supp. 2d at 1235-1239 (discussing the cases that have examined the scope of the FDA's authority with respect to compounding).

[115] See Gonzales v. Raich, 545 U.S. 1, 22 (2005) (holding that the authority under the Commerce Clause extends such that Congress can regulate activities when, taken in the aggregate, substantially affect interstate commerce).

[116] But see Franck's Lab, 816 F. Supp. 2d at 1214 ("[T]he FDA has taken the bright-line position that any compounding of animal medications from bulk substances violates ... the FDCA ...")

[117] See CPG §460.200 (May 29, 2002); see also CPG §608.400 (July 14, 2003).

[118] This lack of resolution could be a considerable burden on traditional compounders, whose conduct could be in violation of federal law. As one court noted, "it remains no small burden for compounding pharmacists ... to 'live in sin,' [as] their livelihood [has] no greater assurance than the FDA's good graces." See Medical Ctr. Pharm., 536 F.3d at 399-400.

[119] See, e.g., Wedgewood Vill. Pharm., 421 F.3d at 272-73.

In: Drug Compounding Pharmacies
Editor: Seth Sharpe

ISBN: 978-1-62808-176-3

Chapter 2

FDA's Authority to Regulate Drug Compounding: A Legal Analysis*

Jennifer Staman

Summary

In light of the recent meningitis outbreak, believed to have been caused by a contaminated compounded steroid injection, the regulation of drug compounding has received significant attention. Compounding is traditionally defined as a process of combining, mixing, or altering ingredients in order to create a medication for a particular patient. However, as in the case of the pharmacy that produced the steroid medication, concerns have been raised about compounding pharmacies producing drugs on a larger scale, something more akin to drug manufacturing. While drug compounding has historically been regulated primarily by states through their regulation of pharmacies, recent questions have been raised about the extent to which the Federal Food, Drug, and Cosmetic Act (FFDCA) governs this practice, and what authority the U.S. Food and Drug Administration (FDA) has to regulate a compounded drug as a "new drug," subject to approval by the FDA, as well as other requirements.

In 1997, Congress enacted the FDA Modernization Act of 1997 (FDAMA), which was a comprehensive revision of the FFDCA. Section

---

* This is an edited, reformatted and augmented version of a Congressional Research Service publication, CRS Report for Congress R40503, prepared for Members and Committees of Congress, from www.crs.gov, dated October 17, 2012.

127 of FDAMA added Section 503A to the FFDCA, which excepted compounded drugs from various "new drug" requirements, conditioned upon the compounded drugs meeting a variety of restrictions. One of the restrictions in Section 503A of the FFDCA was that drug providers were prohibited from soliciting or advertising particular compounded drugs. These speech restrictions were challenged on First Amendment grounds and were struck down by the Supreme Court in Thompson v. Western States Medical Center. Following this decision, there was controversy over the current status of compounded drugs under the FFDCA and whether the remaining provisions of Section 503A remain good law, an issue that the Supreme Court did not address in Western States. The two circuits that addressed this issue took different positions. While the Ninth Circuit in Western States determined that Section 503A was struck down in its entirety, the Fifth Circuit in Medical Center Pharmacy v. Mukasey found that the lawful provisions of Section 503A are still in effect. Accordingly, these cases have created an interesting scenario of non-uniform enforcement throughout the U.S. In the Fifth Circuit, compounded drugs are specifically exempted from new-drug, adulteration, and misbranding requirements of the FFDCA if certain criteria are met; while in the Ninth Circuit (and, according to the FDA, the rest of the United States), compounded drugs are subject to these requirements, but the FDA may exercise discretion in taking action against an entity that violates these provisions. This report provides a brief historical overview of the FDA's regulation of drug compounding and addresses these conflicting decisions. The report will also address the FDA's current authority to regulate compounded drugs under the FFDCA in light of these decisions.

BACKGROUND

Under its traditional definition, drug compounding is a process in which a pharmacist combines, mixes, or alters various drug ingredients to create a medication for an individual patient in response to a practitioner's prescription.[1] It is generally used to prepare medications that are not typically commercially available, such as a drug in a lower dosage for a child, or a drug without a dye or a preservative in response to a patient allergy.[2] Compounding is considered a conventional component of the practice of pharmacy and has historically been regulated at the state level.[3]

The Federal Food, Drug, and Cosmetic Act (FFDCA), as amended, establishes certain minimum standards for the manufacturing, marketing, and distribution of drugs, and authorizes the FDA to ensure that drugs and other products marketed in the United States are safe and effective for their intended uses.[4] The act prohibits any person from introducing a "new drug" into interstate commerce unless it is approved by the FDA. A "new drug" is defined by the act as "[a]ny drug (except a new animal drug or an animal feed bearing or containing a new animal drug) the composition of which is such that such drug is not generally recognized ... as safe and effective for use under the condition prescribed, recommended, or suggested in the labeling thereof."[5] In order to be determined to be safe and effective, new drugs are subject to a lengthy approval process, which is supervised by the FDA. The FFDCA also contains requirements preventing drugs from being adulterated[6] or misbranded.[7]

Until 1997 with the enactment of the FDA Modernization Act of 1997 (FDAMA, discussed below),[8] Congress had not explicitly addressed whether compounded drugs had to meet these requirements of the FFDCA.[9]

For approximately the first 50 years of the FFDCA, pharmacists compounded drugs without getting FDA approval for these drugs, and regulation of drug compounding was left to the states.[10] The reason for this, in large part, is because it is widely recognized that compounded drugs could not meet the FFDCA's drug approval requirements because compounded drugs are traditionally made in small amounts for an individual patient (e.g., safety and efficacy trials are impracticable, and compounding pharmacies cannot afford testing for the new drug approval process to meet the needs of individual patients).[11]

Thus, the question of whether the FDA could regulate compounded drugs was generally not disputed until around the early 1990s when the FDA became concerned that some pharmacists were engaged in large-scale bulk compounding that was, in the FDA's view, more akin to drug manufacturing and an attempt to avoid the FFDCA's new drug requirements.[12] In response to these perceived abuses, the FDA issued a Compliance Policy Guide in 1992 addressing its position regarding its authority to regulate compounding.[13] The FDA declared "that while retail pharmacies ... are exempted from certain requirements of the [FFDCA], they are not the subject of any general exemption from the new drug, adulteration, or misbranding provisions" of the FFDCA.[14]

While the FDA also indicated that its guidance was not intended to affect a pharmacist's traditional role of compounding drugs extemporaneously in reasonable quantities and pursuant to a valid prescription, the FDA warned that it "may, in the exercise of its enforcement discretion, initiate federal enforcement actions ... when the scope and nature of a pharmacy's activity raises the kind of concerns normally associated with a manufacturer." The 1992 Compliance Policy Guide set forth certain factors the FDA would consider in determining whether to exercise enforcement discretion against a compounding pharmacy whose activities raise concerns similar to those of drug manufacturers, including soliciting business to compound specific drug products, using commercial scale manufacturing or testing equipment to compound drug products, and compounding (on a regular basis or in excessive amounts) drug products that are commercially available in the market place and that are essentially generic copies of commercially available, FDA-approved drug products.[15]

In order to "clarify the status of pharmacy compounding under Federal law,"[16] Congress added Section 503A (21 U.S.C. § 353a) to the FFDCA as part of the FDA Modernization Act of 1997 (FDAMA), which specifically addressed the FDA's role in the regulation of drug compounding. Section 503A of the FFDCA, as added by FDAMA, exempts compounded drugs from FFDCA requirements regarding drug adulteration, misbranding, and new drug approval, provided that certain conditions are satisfied.[17]

Among these requirements, a drug product must be compounded by a licensed pharmacist or physician for an indentified individual patient based on a valid prescription.[18]

The compounded drug must comply with standards of an applicable U.S. Pharmacopoeia, or made from FDA-approved drug ingredients, meet certain manufacturing criteria, and the drug compounded must not be one that appears on a list of drugs (published by the Secretary) of drug products that have been withdrawn or removed from the market because the product, or components of the product have been found to be unsafe or not effective.[19] Further, the drug provider compounding the drug may not "compound regularly or in inordinate amounts ... any drug products that are essentially copies of a commercially available drug product."[20]

Section 503A of the FFDCA also provided that the exemptions from the FFDCA's new drug and other requirements only applied to drug products compounded based on a valid prescription that was not solicited.[21] Further, the section states that a drug may be compounded and subject to the exemptions only if the pharmacy, licensed pharmacist, or licensed physician does not advertise or promote the compounding of any particular drug.[22] In November of 1998, just days after the provision became effective, several compounding pharmacies challenged the restrictions on solicitation and advertising as being an impermissible regulation of speech under the First Amendment.

THOMPSON V. WESTERN STATES MEDICAL CENTER[23]

In *Western States*, a group of licensed pharmacies brought an action against the Secretary of Health and Human Services (HHS) and the Commissioner of the FDA, claiming that FDAMA's restrictions on solicitation and advertising violated the pharmacies' rights to free speech under the First Amendment.[24] The district court evaluated whether the provisions were an impermissible government regulation of commercial speech[25] under the factors articulated in *Central Hudson Gas and & Electric v. Pub. Service Comm'n of New York*,[26] and granted the pharmacies summary judgment on the basis that the provision was indeed a violation of the First Amendment.

In evaluating whether the other provisions of Section 503A remained intact, the district court found that there is "no question" that the other sections could be severed[27] from the remaining sections.[28] The court explained that the modification of the statute did not prevent it from operating as law, based on the numerous other requirements that must be met in the production and distribution of compounded drugs. The Ninth Circuit Court of Appeals affirmed the district court's holding that the speech provisions of section 503A were unconstitutional, but also reversed in part, holding that the compounding provisions were not severable from the rest of the section.[29]

In discussing the severability issue, the court explained that the solicitation and advertising provisions were an essential component in balancing the preservation of compounded drugs alongside the desire to prevent pharmacists and others from engaging in drug manufacturing. The court pointed to legislative history that, in its opinion, demonstrated an intent to strike this balance.[30] The court further indicated that the legislative record demonstrated that Congress meant to exempt compounding pharmacists from FFDCA requirements only in return for a prohibition on the promotion of specific compounded drugs. The court also dismissed the applicability of the existing severability clause in the FFDCA[31] as not indicative of a presumption that Congress intended for unoffending provisions of FDAMA to remain valid. The court explained that because Congress approved the severability clause before FDAMA's passage, there was no indication of whether the clause was to only apply to the original FFDCA, or to the FFDCA as amended by FDAMA.

In its brief requesting Supreme Court review on the First Amendment issue, the government indicated its acquiescence with the court of appeals' conclusion that, if the solicitation and advertising provisions in Sections 503A(a) and (c) are unconstitutional, they are not severable from the other provisions of Section 503A. Accordingly, the government did not seek review of that holding.[32] The Supreme Court later affirmed the lower court rulings that the speech restrictions were unconstitutional, but because neither party petitioned for certiorari on the severability issue, the Court articulated that it "had no occasion" to review that portion of the court of appeals' decision.[33]

2002 FDA Compliance Policy Guide

Western States did not address the FDA's authority to regulate drug compounding in the absence of the FDAMA provision. Because (at least in part) of the confusion about the extent of the FDA's authority to regulate compounding remaining after this decision, the FDA issued a Compliance Policy Guide in 2002, largely similar to the compliance guide issued in 1992.[34] The FDA stated that because the Supreme Court did not rule on the severability issue, the Court "therefore left in place" the Ninth Circuit holding that the other provisions of Section 503A were not severable, and the section was therefore invalid. Accordingly, the FDA determined that it needed to issue guidance to the compounding industry regarding the factors the agency would consider in exercising enforcement discretion with regard to pharmacy compounding.

The FDA indicated that it did not intend the Compliance Policy Guide to regulate traditional compounding practices (i.e., pharmacists who extemporaneously compound reasonable quantities of drugs upon receipt of a valid prescription for an identified patient). In addition, the FDA noted that it would continue to defer to the state authorities regarding "less significant violations" of the FFDCA. However, the FDA explained its concern with pharmacies that were not compounding in the regular course of business, but were manufacturing large quantities of drug products without a prescription for them. The FDA explained that when the scope and nature of a pharmacy's activities raise the kinds of concerns normally associated with a drug manufacturer and result in significant violations of the new drug, adulteration, or misbranding provisions of the act, it determined that it should seriously consider enforcement action. The FDA set out nine acts it would consider in pursuing an enforcement action:

1. Compounding of drugs in anticipation of receiving prescriptions, except in very limited quantities in relation to the amounts of drugs compounded after receiving valid prescriptions

2. Compounding drugs that were withdrawn or removed from the market for safety reasons
3. Compounding finished drugs from bulk active ingredients that are not components of FDA-approved drugs, without an FDA sanctioned investigational new drug application
4. Receiving, storing, or using drug substances without first obtaining written assurance from the supplier that each lot has been made in an FDA-registered facility
5. Receiving, storing, or using drug components not guaranteed to meet official compendia requirements
6. Using commercial scale manufacturing or testing equipment for compounding drug products
7. Compounding drugs for third parties who resell to individual patients or offering compounded drug products at wholesale to other state licensed persons or other commercial entities for resale
8. Compounding drug products that are commercially available in the marketplace or that are essentially copies of commercially available FDA-approved drug products
9. Failing to operate in conformance with applicable state law regulating the practice of pharmacy.

The FDA indicated that the list of acts was not exhaustive, and that other factors may be appropriate to consider in enforcement actions.

MEDICAL CENTER PHARMACY V. MUKASEY

In September 2004, 10 licensed pharmacies specializing in drug compounding brought suit challenging FDA's authority to regulate compounded drugs.[35] In *Medical Center Pharmacy v. Ashcroft* (later changed to *Medical Center Pharmacy v. Gonzalez* and then *Medical Center Pharmacy v. Mukasey*), plaintiffs sought, among other things, a declaration that drugs compounded by licensed pharmacists were not "new drugs" per se under the FFDCA. The district court granted summary judgment on the pharmacies' claim, finding that compounded drugs were "implicitly exempt" from the new-drug requirements.[36] The court, relying on language of the Supreme Court from *Western States*, reasoned that if compounded drugs were not exempt from the new-drug requirements, the drugs would have to undergo the new-drug approval process, which would be impracticable. The court indicated that the FDA's 2002 Compliance Policy Guide, which exempts traditional compounding from the scope of the guidance, and instead focuses on the regulation of pharmacies that manufacture drugs "under the guise of compounding," provided additional support for the position that traditional compounding was not subject to the new-drug approval requirements. The court also opined that the remainder of Section 503A was severable following *Western States*. The court explained that it was not bound by the Ninth Circuit, and that the severability statute in the FFDCA is "clear and unambiguous."[37]

On appeal, the Fifth Circuit disagreed with the district court, finding that compounded drugs are "new drugs" and are subject to the drug approval, adulteration, and misbranding requirements.[38] However, in light of the fact that the provisions of 503A were severable, the court found that compounded drugs are subject to a "limited exemption" from the those requirements if the drugs are compounded in accordance with the non-speech provisions of that section. The court based its holding in part on the definition of "new drug," under the FFDCA. Given that a compounded drug is created by mixing or combining an approved drug with something else to create a "concoction" that has not been previously approved for use, a compounded drug was, based on the plain language of the statute, a new drug.[39] Further, given that the FFDCA contains certain exceptions to the new drug definition (e.g., for "grandfathered" drugs), the court found meaning in the lack of an exemption to the new-drug definition for drugs created by compounding.

The pharmacies in *Medical Center Pharmacy* had argued that including compounded drugs under the new drug definition would, in effect, extinguish the practice of traditional compounding. The court also acknowledged the idea that interpreting the new-drug definition so as to include compounding did appear inconsistent with the idea that compounding should be able to persist and not be subject to the rigorous requirements of new drug approval. However, the court pointed out that these issues can be reconciled. Under the court's reasoning, even if compounded drugs are effectively made unlawful by the new drug definition and approval requirements, pharmacists could still continue compounding because the FDA has the ability exercise enforcement discretion and can decline enforcement of "minor violations."[40] In other words, the FDA would be in a position, as it did historically, to use this discretion and not enforce new drug requirements on compounded drugs. While the court expressed some reluctance to rely on the FDA's discretion in its enforcement of drug compounding, the court still found that because the FDA has demonstrated willingness to accommodate traditional compounding, pharmacies would still be able to compound drugs even if they were deemed new drugs.

The Fifth Circuit also pointed to the enactment of FDAMA as evidence that Congress intended that the new-drug provision apply to drugs created by pharmacy compounding.[41] However, in order to rely on FDAMA as justification for why compounded drugs could be considered new drugs under the FFDCA, the court evaluated whether the statute was still valid in light of the *Western States* decision. The court examined the severability clause in the FFDCA and found, contrary to the Ninth Circuit, that the severability clause applied to Section 503A. The court explained that if Congress did not want the FFDCA's severability clause to apply to Section 503A, it would have specifically said so. Further, the court did not see the free speech provisions as so central to the purpose of FDAMA that Congress would not have passed the statute without them. The Fifth Circuit also stated that the remaining requirements were sufficient to accomplish the goal of protecting access to compounded drugs while preventing pharmacies from engaging in large-scale drug manufacturing.[42] The court also disagreed with the Ninth Circuit in finding that the legislative history to FDAMA was "inconclusive" as to whether Congress would not have enacted FDAMA without the speech provisions of Section 503A. 43

ISSUES WITH CURRENT FDA AUTHORITY TO REGULATE DRUG COMPOUNDING

As discussed above, the *Western States* and *Medical Center Pharmacy* cases directly conflict on the question of whether the non-speech provisions of Section 503A are severable and thus, whether the remaining sections of the statute are still in effect. Based on these two decisions, the FDA has determined that it will apply the non-advertising provisions of Section 503A of the FFDCA to entities that are located within the jurisdiction of the Fifth Circuit (Texas, Louisiana, and Mississippi) as well as to the plaintiffs that brought the *Medical Center* case.[44] In all other locations, the FDA will continue to apply the enforcement policy articulated in the 2002 Compliance Policy Guide. In other words, the current situation presents an interesting scenario of non-uniform enforcement throughout the United States. In the Fifth Circuit, compounded drugs are specifically exempted from adulteration, misbranding, and new drug provisions of the FFDCA, subject to meeting certain criteria; while in the Ninth Circuit (and, according to the FDA, the rest of the United States), compounded drugs are subject to these provisions, but the FDA may exercise discretion in taking action against a pharmacy that violates them.

The parties in *Medical Center Pharmacy* did not petition the Supreme Court for review, and thus, uncertainty remains about the FDA's authority to regulate compounded drugs as new drugs. Courts may still evaluate the FDA's position that compounded drugs are new drugs, or, relatedly, whether the non-speech provisions of Section 503A of the FFDCA that include an exemption for these drugs remain intact. If these types of challenges were to occur outside of the Fifth or Ninth Circuits, a court could rely upon, but would be under no obligation to follow, the *Western States/Medical Center Pharmacy* precedent. Unless a future case on these issues reaches the Supreme Court, the question of how the FDA may regulate compounded drugs may not be settled unless Congress takes legislative action.

Under the current drug compounding regime outside of the Fifth Circuit, some have expressed concern that the FDA's 2002 Compliance Policy Guide creates confusion regarding when FDA enforcement authority will be used.[45] It has been argued that the FDA Compliance Policy Guide lacks a clear description of the circumstances under which the agency will take action against pharmacies.[46] The FDA has countered that the guidance gives the agency flexibility in responding a wide variety of situations involving public health and safety issues.[47]

The FDA's authority to regulate compounded drugs is set forth in a guidance document. In contrast to agency rules, which have the force and effect of law, guidance documents are merely considered to be a general statement of policy.[48] These statements of policy are agency statements that "advise the public prospectively of the manner in which the agency proposes to exercise a discretionary power."[49] Congress has passed requirements specific to FDA guidance documents, which state that such documents "shall not create or confer any rights for or on any person, although they present the views of the Secretary on matters under the jurisdiction of the Food and Drug Administration."[50] Under regulations prescribing FDA good guidance practices, it is stated that guidance documents do not establish legally enforceable rights or responsibilities and do not legally bind the public or the FDA."[51] Accordingly, if a court were to evaluate the position the FDA takes in its 2002 Compliance Policy Guide, a court may choose not to defer to the FDA's position on when it may exercise enforcement discretion, or whether compounded drugs are "new drugs." However, in *Medical Center Pharmacy*, the Fifth Circuit evaluated the FDA's statutory interpretation that compounded drugs were new drugs under a standard that is more deferential to agency action, a standard that is typically reserved for products of formal agency process (e.g., notice and comment rulemaking). [52] It should be noted that the few courts that have evaluated the Compliance Policy Guide have found the guidance to be appropriate or reasonable.[53]

If a future court were to take up the issue of whether the non-speech provisions in Section 503A are severable, it is difficult to predict how a court would rule. Under Supreme Court severability precedent, "[u]nless it is evident that the Legislature would not have enacted those provisions which are within its power, independently of that which is not, the invalid part may be dropped if what is left is fully operative as a law."[54] Further, a key inquiry evaluating severability is whether without the unconstitutional provision, "the statute will function in a manner consistent with the intent of Congress."[55] The Court has also indicated "when Congress has explicitly provided for severance by including a severability clause in the statute ... the inclusion of such a clause creates a presumption that Congress did not intend the validity of the statute in question to depend on the validity of the constitutionally offensive provision."[56] The courts in both *Western States* and *Medical Center Pharmacy* relied on various statements in the legislative history of FDAMA as evidence of whether the solicitation and advertising provisions of FDAMA were integral to the rest of the statute. However, neither court was able to point to a statement that explicitly referenced the unconstitutional provisions of Section 503A and the intent behind them. In addition, as in these two cases, if courts in another jurisdiction were to review whether Section 503A is severable, a court would likely evaluate the severability clause of the FFDCA. A court could agree with the Ninth Circuit, finding that because the severability clause was enacted prior to FDAMA, it is not clear whether it was to apply to the amended portions of the FFDCA.[57] Alternatively, a court could side with the Fifth Circuit and hold that the severability clause kept Section 503A of the FFDCA intact, assuming that if Congress had not wanted the clause to apply, it would have said so.[58]

End Notes

[1] See Western States Medical Center v. Shalala, 238 F.3d 1090, 1092 (9th Cir. 2001).

[2] Western States Medical Center et. al., v. Shalala, 69 F. Supp. 2d 1288, 1292 (D. Nev. 1999). See also Statement of Steven K. Galson, Acting Director, Center for Drug Evaluation and Research, U.S. Food and Drug Administration, before the Senate Committee on Health,

Education, Labor and Pensions, Hearing on "Federal and State Role in Pharmacy Compounding and Reconstitution: Exploring the Right Mix to Protect Patients," (Oct. 23, 2003).

[3] It should be noted that drugs may be compounded for animal use. See, e.g., Medical Center Pharmacy v. Mukasey, 536 F.3d 383, 406-409 (5th Cir. 2008); United States v. Algon Chemical, Inc., 879 F.2d 1154 (3rd Cir. 1989). The statutes, regulations, and FDA policies governing compounding for animals differ from those governing drug compounding for humans. This report will only address federal law as it relates to drug compounding for human use.

[4] 21 U.S.C. § 301 et seq. The FFDCA regulates several other products besides drugs, including certain foods, cosmetics, and dietary supplements.

[5] 21 U.S.C. § 321(p). Exceptions are made for certain "grandfathered" drugs used prior to the FFDCA, as well as drugs intended only for investigational use.

[6] Under the FFDCA, a drug may be deemed adulterated if "the methods used in, or the facilities or controls used for, its manufacture, processing, packing, or holding do not conform to or are not operated or administered in conformity with current good manufacturing practice to assure that such drug meets the requirements of this Act ... as to safety and has the identity and strength, and meets the quality and purity characteristics, which it purports or is represented to possess." See 21 U.S.C. § 351 for a complete definition. FDA regulations set forth various standards for good manufacturing practices, including standards relating to personnel, facilities, and drug product containers. 21 C.F.R. § 211.1 et seq.

[7] The FDA approves drugs for specific uses that are reflected in their labeling. A drug may be deemed to be misbranded if, among other possibilities, the labeling is false or misleading or if its labeling does not bear "adequate directions for use." 21 U.S.C. § 352.

[8] Food and Drug Administration Modernization Act, P.L. 105-115, 111 Stat. 2296 (Nov. 21, 1997).

[9] It should be noted that the FFDCA explicitly addresses compounded drugs in at least two notable areas. First, under Section 510(g) of the FFDCA, pharmacies are exempt from registering as drug manufacturers under the act if the pharmacies maintain establishments "in conformance with any applicable local laws regulating the practice of pharmacy and medicine and which are regularly engaged in dispensing prescription drugs or devices, upon prescriptions of practitioners licensed to administer such drugs or devices to patients under the care of such practitioners in the course of their professional practice, and which do not manufacture, prepare, propagate, compound, or process drugs or devices for sale other than in the regular course of their business of dispensing or selling drugs or devices at retail." 21 U.S.C. § 360(g)(1). In other words, this provision appears to recognize that traditional compounding by pharmacists is not manufacturing. Richard R. Abood, PHARMACY PRACTICE AND THE LAW, Fifth Ed. (2008). Second, similar language limits the types of pharmacy records that the FDA may review during an inspection. See 21 U.S.C. § 374.

[10] See Western States, 535 U.S. at 362.

[11] Statement of Steven K. Galson, Acting Director, Center for Drug Evaluation and Research, U.S. Food and Drug Administration, before the Senate Committee on Health, Education, Labor and Pensions, Hearing on "Federal and State Role in Pharmacy Compounding and Reconstitution: Exploring the Right Mix to Protect Patients," (Oct. 23, 2003). See also Thompson v. Western States, 535 U.S. 357, 369 (2002) ("...it would not make sense to require compounded drugs created to meet the unique needs of patients to undergo the testing required for the new drug approval process. Pharmacists do not make enough money from small-scale compounding to make safety and efficacy testing of their compounded

drugs economically feasible, so requiring such testing would force pharmacists to stop providing compounded drugs.")

[12] Medical Center Pharmacy, 536 F.3d at 389. The FDA's concern over the practice of compounding has also been, in large part, due to certain patient injuries caused by compounded drugs. The Agency has asserted that it knows of more than 200 adverse events involving 71 compounded products since 1990, including certain instances with "devastating repercussions." See, U.S. Food and Drug Administration, The Special Risks of Pharmacy Compounding, available at http://www.fda.gov/consumer/updates/ compounding 053107.html.

[13] Compliance Policy Guide No. 7132.16 (Mar. 1992).

[14] Id., as cited in Western States, 535 U.S. at 362. It should also be noted that prior to the enactment of FDAMA, courts found that compounded drugs were new drugs. See, e.g., Prof'ls & Patients for Customized Care v. Shalala, 56 F.3d 592, 593 n.3 (5th Cir. 1995).

[15] Compliance Guide 7132.16, as cited in Medical Center Pharmacy, 536 F.3d at 391.

[16] Compliance Policy Guide 460.200 (May 2002); see also H.Rept. 105-399 at 94, which addresses section 503A of the FFDCA ("It is the intent of the conferees to ensure continued availability of compounded drug products as a component of individualized therapy, while limiting the scope of compounding so as to prevent manufacturing under the guise of compounding. Section 503A establishes parameters under which compounding is appropriate and lawful.").

[17] See Section 127 of P.L. 105-115.

[18] 21 U.S.C. § 353(a)

[19] 21 U.S.C. § 353a(b)(1).

[20] 21 U.S.C. § 353a(b)(1)(D).

[21] 21 U.S.C. § 353a(a).

[22] 21 U.S.C. § 353a(c).

[23] Thompson v. Western States Medical Center, 535 U.S. 357 (2002).

[24] Western States Medical Center, et al., v. Shalala, 69 F. Supp. 2d 1288 (D. Nev. 1999).

[25] The First Amendment to the United States Constitution provides that "Congress shall make no law ... abridging the freedom of speech, or of the press.... " Despite its absolute language, it provides no protection to some types of speech and only limited protection to others. One type of speech to which it applies only limited protection is commercial speech, which is "speech that proposes a commercial transaction." Board of Trustees of the State University of New York v. Fox, 492 U.S. 469, 482 (1989). Commercial speech may be banned if it advertises an illegal product or service, and, unlike fully protected speech, may be banned if it is unfair or deceptive. For a more general discussion of First Amendment jurisprudence regarding freedom of speech, see CRS Report 95-815, Freedom of Speech and Press: Exceptions to the First Amendment, by Kathleen Ann Ruane.

[26] 447 U.S. 557 (1980). For commercial speech, the Supreme Court has prescribed the four-prong Central Hudson test to determine its constitutionality. This test asks initially (1) whether the commercial speech at issue is protected by the First Amendment (that is, whether it concerns a lawful activity and is not misleading) and (2) whether the asserted governmental interest in restricting it is substantial. "If both inquiries yield positive answers," then to be constitutional the restriction must (3) "directly advance[ ] the governmental interest asserted," and (4) be "not more extensive than is necessary to serve that interest." Id. at 566 (1980).

[27] When one section of a law is held unconstitutional, courts are faced with determining whether the remainder of the statute is "severable," i.e., it remains valid, or whether the whole

statute is nullified. In general, courts examine whether the statute can fully operate without the unconstitutional provision in the manner that Congress intended. See infra notes 54-58 and accompanying text.

[28] Western States, 69 F. Supp. 2d at 1309-10.

[29] Western States Medical Center v. Shalala, 238 F.3d 1090 (9th Cir. 2001).

[30] For example, the Ninth Circuit cited a House report explaining that FDAMA was designed to "ensure continued availability of compounded drug products as a component of individualized drug therapy, while limiting the scope of compounding so as to prevent manufacturing under the guise of compounding." H.Rept. 105-399, as cited in Western States, 238 F.3d at 1097.

[31] Section 901 of the FFDCA provides, "If any provision of this Act [21 U.S.C. §§ 301 et. seq.] is declared unconstitutional, or the applicability thereof to any person or circumstances is held invalid, the constitutionality of the remainder of the Act ... and the applicability thereof to other persons and circumstances shall not be affected thereby." 21 U.S.C. § 391.

[32] Thompson v. Western States Med. Ctr., 2001 U.S. Briefs 344, 11 (Aug. 24, 2001)

[33] Western States, 535 U.S. at 365.

[34] Food and Drug Administration, Compliance Policy Guides Manual § 460.200, Pharmacy Compounding (May 2002), available at http://www.fda.gov/ora/compliance_ref/cpg/cpgdrg/cpg460-200.html#BACKGROUND.

[35] Medical Center Pharmacy, et al., v. Gonzales, et al., 451 F. Supp. 2d 854 (W.D. Tex.)(2006).

[36] Id. at 858.

[37] Id. at 864.

[38] Medical Center Pharmacy v. Mukasey, 536 F.3d 383 (5th Cir. 2008).

[39] Id. at 394-95.

[40] Id. at 398-99.

[41] Id. at 400.

[42] The Fifth Circuit also pointed to the Supreme Court's evaluation of the advertising provision in Western States:

Several non-speech-related means of drawing a line between compounding and large-scale manufacturing might be possible here.... It might even be sufficient to rely solely on the nonspeech-related provisions of FDAMA, such as the requirement that compounding only be conducted in response to a prescription or a history of receiving a prescription, 21 U.S.C. § 353a(a), and the limitation on the percentage of a pharmacy's total sales that out-of-state sales of compounded drugs may represent, § 353a(b)(3)(B).... Nowhere in the legislative history of FDAMA or petitioners' briefs is there any explanation of why the Government believed forbidding advertising was a necessary as opposed to merely convenient means of achieving its interests.

Medical Ctr. Pharm. v. Mukasey, 536 F.3d at 404 n.43 (citing Western States, 535 U.S. at 372-73.) (emphasis added).

[43] See also United States v. Bader, 2009 U.S. Dist. LEXIS 63490 (D. Colo. 2009), where a Tenth Circuit district court evaluated a motion for reconsideration over whether human growth hormone imported by a pharmacist was a "finished, consumer-usable" drug that was improperly smuggled and distributed under various federal criminal provisions, and the FFCDA, or whether it was used as an ingredient in a compounded drug under which the pharmacist may be able to escape liability. In discussing whether the pharmacist had engaged in drug compounding, because Bader was a party in the Medical Center Pharmacy case, the court found that the provisions of FDAMA were considered severable. The court,

in granting the motion, found that the trial court would need to determine, among other things, whether the pharmacist's conduct violated the terms of FDAMA.

[44] See, e.g., Letter from Alonza E. Cruse, Director, Los Angeles District, U.S. Food and Drug Administration, to Ari S. Schafer, Civic Center Pharmacy (Dec. 16, 2008), available at http://www.fda.gov/foi/warning_letters/s7071c.pdf.

[45] U.S. Government Accountability Office, GAO-04-195T Prescription Drugs, State and Federal Oversight of Drug Compounding by Pharmacies, Testimony Before the Committee on Health, Education, Labor, and Pensions, U.S. Senate (Oct. 2003) at 12.

[46] Id. According to GAO testimony, pharmacy associations pointed to terms in the guidance, such as "very limited quantities" and "commercial scale manufacturing or testing equipment" that are not defined, and observed that FDA also has the right to consider other factors in addition to those in the guidance without giving further clarification. Id. at 12.

[47] Government Accountability Office, note 45 supra.

[48] Jeffrey S. Lubbers, A GUIDE TO FEDERAL AGENCY RULEMAKING 94 (4th ed. 2006). Executive Order 13422 sets forth a definition of a guidance document. Executive Order 13422, Further Amendment to Executive Order 12866 on Regulatory Planning and Review, Jan. 18, 2007, http://www.whitehouse.gov/news/releases/2007/01/20070118.html.

[49] Tom C. Clark, Attorney General, Attorney General's Manual on the Administrative Procedure Act, at 30 n.3 (1947), http://www.law.fsu.edu/library/admin/1947iii.html; see, e.g., Chamber of Commerce v. United States Department of Labor, 174 F.3d 206, 212 (D.C. Cir. 1999).

[50] 21 U.S.C. § 371(h). The provisions on guidance documents were added in 1997 by FDAMA § 405.

[51] 21 C.F.R. § 10.115(d)(1).

[52] Under Supreme Court precedent, certain agency actions should be afforded deference by the courts. In Chevron U.S.A. Inc. v. Natural Resources Defense Council, 467 U.S. 837 (1984), the Supreme Court established a two-part test for judicial review of agency statutory interpretations. First, a reviewing court must determine "whether Congress has directly spoken to the precise question at issue." If it has, the agency must of course comply with clear congressional intent, and regulations to the contrary will be invalidated. Second, in instances where congressional intent is not clear and the statutory language is ambiguous, the courts will likely defer to any reasonable agency interpretation, even if another interpretation is more plausible. However, the Supreme Court in Christensen v. Harris County, 529 U.S. 576 (2000) and United States v. Mead Corp., 533 U.S. 218 (2001) analyzed Chevron's application, ruling that Chevron deference applies only if an agency's interpretation is the product of a formal agency process, such as adjudication or notice-and-comment rulemaking, through which Congress has authorized the agency "to speak with the force of law." Mead Corp., 533 U.S. at 229. Although a notice about the Compliance Policy Guide was published in the Federal Register and public comments were requested (in accordance with the FDA's good guidance practices), it would seem more likely that a court would not employ Chevron deference in evaluating the agency's interpretation as provided in the guidance document. But, as discussed above, courts still employ this deferential standard outside of this formal agency action. See also In re the Matter of Establishment Inspection of Wedgewood Pharmacy Inc., 270 F. Supp. 2d 525 (D.N.J. 2003) (court applies Chevron deference to 2002 Compliance Policy Guide).

[53] See, e.g., In the Matter of Establishment Inspection of Wedgewood Pharmacy, 421 F.3d 263 (3rd Cir. 2005)(court of appeals finds the guide "a reasonable basis under which to initiate an inspection under the FFDCA"). See also Schaerrer v. Stewart's Plaza Pharm., Inc., 79

P.3d 922, (Utah 2003)(In examining whether compounding pharmacy's conduct fell into category of "pharmacist" or "drug manufacturer," court states that "[w]hile the policy statements of the FDA are by no means binding on this court, they do provide meaningful guidance on a question that few, if any, courts in this country have yet considered."). It should also be noted that FDA regulations define a guidance document to include "documents that relate to: ... inspection and enforcement policies." 21 C.F.R. § 10.115(b)(2). Thus, while not addressed in the Wedgewood case, given that the Compliance Policy Guide addressed when the FDA will exercise enforcement discretion on drug compounding, one could argue that the guide is an appropriate exercise of FDA authority.

[54] Alaska Airlines, Inc. v. Brock, 480 U.S. 678, 684 (1987)(quoting Buckley v. Valeo, 424 U.S. 1, 108 (1976) (per curiam), Champlin Refining Co. v. Corporation Comm'n of Oklahoma, 286 U.S. 210, 234 (1932).

[55] Alaska Airlines, 480 U.S. at 685.

[56] Id. at 686 (citing INS v. Chadha, 462 U.S. 919, 932 (1983); Champlin Refining Co. v. Corporation Comm'n of Oklahoma, 286 U.S. 210, 235 (1932).

[57] Western States, 238 F.3d at 1098-99.

[58] Medical Center Pharmacy, 536 F.3d at 401-402.

In: Drug Compounding Pharmacies
Editor: Seth Sharpe
ISBN: 978-1-62808-176-3

Chapter 3

STATE OF DISARRAY: HOW STATES' INABILITY TO OVERSEE COMPOUNDING PHARMACIES PUTS PUBLIC HEALTH AT RISK*

Edward J. Markey

EXECUTIVE SUMMARY AND RESULTS IN BRIEF

An outbreak of fungal meningitis that has thus far claimed the lives of 53 people and sickened 733 brought national attention to compounding pharmacies, the drugs they produce, and the potential risks from these drugs. At the center of this outbreak was the Massachusetts-based New England Compounding Center (NECC) that produced the preservative-free injectable steroid drug that caused the deadly outbreak. This drug, found later to be contaminated with mold, was shipped to 23 states and injected into more than 14,000 patients, resulting in the worst pharmaceutical-related public health crisis in U.S. history. The NECC tragedy was not the first incident of compounding pharmacies causing fatalities or illnesses. Over the last decade, compounding pharmacies have been responsible for at least 23 additional deaths and 86 serious illnesses across the nation.

---

* This is an edited, reformatted and augmented version of the U.S. House of Representatives, dated April 15, 2013.

Unlike drug manufacturers, which are subjected to rigorous Food and Drug Administration (FDA) oversight, there is no requirement for compounding pharmacies to register with the FDA or to follow any FDA-specified procedures to ensure that the drugs they produce are safe and effective. In fact, the oversight of all pharmacies typically falls under the purview of individual states. Attempts by the FDA and by Congress to address safety concerns by ensuring uniform federal standards for compounding pharmacies have led to numerous lawsuits and conflicting judicial rulings and according to the FDA, created "ambiguity associated with FDA's authority over compounding pharmacies." The agency has called on Congress to take action to prevent another such outbreak, stating that if FDA's authority remains unchanged, "this type of incident will happen again. It's a matter of when, not if."

"This type of incident will happen again. It's a matter of when, not if."

-Food and Drug Administration

In November 2012, Reps. Edward J. Markey (D-Mass.), Henry A. Waxman (D- Calif.), John Dingell (D-Mich.), Frank Pallone (D-N.J.) and Diana DeGette (D-Colo.) launched an investigation to examine how state boards of pharmacies oversee compounding pharmacies, requesting detailed information from all boards on their regulatory and information collection processes.[1] This report analyzes the information provided in response to this request. A summary of all state board responses can be found in Appendix B.

The results of this investigation reveal that most states are incapable of assuring the safety of compounded drugs that are prepared using even the riskiest sterile processes or that are shipped into their state from an out-of-state pharmacy. The investigation also found that poor record-keeping practices by the states make it difficult, if not impossible, for the states to identify compounding pharmacies with systemic, repetitive compounding safety problems. Instead of being able to proactively address safety problems, many states rely on public complaints made after a compounded drug is implicated in an injury or death before they can initiate an investigation and take steps to prevent harm to more patients. Moreover, when problems are identified with compounded drugs from pharmacies located in another state, there is no uniform method by which the states notify the FDA or the state in which the problematic pharmacy is located. As a result, these unsafe pharmacies often continue to operate, potentially shipping dangerous or even deadly products across the nation, without increased oversight or penalties.

While most states do not separately track compounding pharmacy activities and the problems associated with them, some states also do a poor job of tracking and overseeing even traditional pharmacy activities. For example, in addition to failing to keep and track records of compounding pharmacy concerns, states including South Dakota, Illinois, Hawaii, New York, Ohio and Vermont, also generally do not routinely track the number of regular pharmacy inspections that are performed. New York, Ohio, Minnesota and Connecticut do not keep searchable records on the number of disciplinary actions that are taken against pharmacies for any reason. The inability of these states to maintain basic records on general pharmacy operations calls into question the capability of these states to oversee more complex and more risky compounding pharmacy activities.

Some states have initiated administrative or legislative improvements to their compounding safety efforts. Despite these efforts, however, deficiencies associated with state record-keeping, communication, resources, and specialized expertise combined with the inability of any single state to monitor the adequacy of compounding pharmacy safety in 49 other states appear to be systemic barriers to relying solely on the states to assure the safety of compounded drug products. The FDA needs clear, unambiguous authority that would enable it to set and enforce safety standards for the riskiest and largest compounding pharmacies, as well as those that sell compounded drugs across state lines.

Finding 1: State boards of pharmacy generally do not know which pharmacies engage in compounding, do not know whether pharmacies ship compounded drugs across state lines, and do not know which pharmacies manufacture large quantities of compounded drugs. In many cases, states are incapable of even providing accurate information regarding the numbers of registered pharmacies in their states.

- Only two states, Mississippi and Missouri, routinely track the number of compounding pharmacies in their state.
- Only thirty-two states were able to provide historical data on the number of licensed pharmacies in their states.
- None of the state boards have requirements for pharmacies to disclose the volumes of compounded drugs they produce or whether compounded drugs are being sold across state lines.

Finding 2: Only thirteen state boards of pharmacy know which pharmacies are providing sterile compounding services and only five of these states have inspectors that are trained to identify problems with sterile compounding.

- Thirty-seven state boards of pharmacy (74% of respondents) do not routinely track which pharmacies are providing risky sterile compounding services
- Only 19 state boards of pharmacy provide their inspectors with special training to identify problems with sterile compounding.

Finding 3: States typically do not maintain pharmacy inspection records that enable them to identify systemic and repeated compounding pharmacy safety problems that originate either in-state or out-of-state.

- Twenty-two state boards of pharmacies do not keep any historical inspection records for compounding pharmacies.

Finding 4: States are unable to effectively police compounding pharmacy activities in other states. Moreover, when issues arise with out-of-state pharmacies, states do not consistently inform the origination state or the FDA.

Finding 5: Despite general increases in state board of pharmacy budgets, the number of pharmacy inspectors has remained consistently low. Furthermore, states usually do not distinguish between inspections of traditional and compounding pharmacies.

- On average, states employ just 5 inspectors (a range of 1-30 inspectors was reported) with responsibility to inspect all pharmacies.
- Only 5 state boards of pharmacy were able to provide an estimation of the number of inspections occurring at compounding pharmacies.

Background

NECC and the Outbreak of Fungal Meningitis

On September 21, 2012 the Centers for Disease Control and Prevention (CDC) was notified by the Tennessee Department of Health (TDH) of a patient who had developed a rare type of fungal meningitis 19 days after receiving an epidural steroid injection at an ambulatory surgical center in Nashville. Less than a week later, the CDC and TDH had discovered several other infected patients and linked the outbreak to the same preservative-free injectable steroid produced by the New England Compounding Center (NECC) in Framingham, Massachusetts.

Once identified as the likely culprit of the infections, NECC initiated a voluntary recall of the suspected lots, identifying more than 17,000 doses shipped to customers in 23 states. By the time the voluntary recall was initiated, more than 14,000 patients had already received a potentially contaminated injection. To date, these tainted drugs have been responsible for 53 deaths and 733 serious illnesses across 20 states.

Following the identification of NECC as the source of the fungal meningitis outbreak, the Massachusetts Department of Public Health and the Food and Drug Administration (FDA) inspected the NECC facility and identified significant problems with cleanliness and sterilization processes at the facility.[2] These problems included observing mold and other foreign material on surfaces of areas in the facility that should have been kept sterile.[3] FDA also found that NECC's own monitoring systems had identified unsafe levels of bacteria and mold, but the company had taken no action.[4]

In addition to the contamination, investigators also found evidence that the NECC had not been compounding drugs for patient-specific prescriptions, as is required of licensed pharmacies under both state and federal regulation. Instead, the NECC accepted patient lists generated by out-patient medical centers, clinics, hospitals, and pain management facilities and provided to NECC for the purpose of obtaining its products. Despite NECC selling contaminated drugs in this manner to at least 23 states, only Colorado had identified NECC as being involved in producing drugs in the absence of a patient-specific prescription two months before the outbreak occurred.[5] At the time of the discovery, the Colorado board notified the Massachusetts board of pharmacy director, who was subsequently fired for never acting on this information.

While this most recent public health nightmare has garnered wide public attention, it isn't the first time that a compounding pharmacy has been at the center of a contamination crisis. On October 29, 2012, Rep. Markey released a report entitled "Compounding Pharmacies, Compounding Risk"[6] that documented more than a decade of violations and problems at compounding pharmacies throughout the country. This report revealed that even before the current meningitis outbreak, compounding pharmacies were responsible for at least 23 deaths and 86 serious illnesses in at least 34 states. Many of the violations at these compounding pharmacies mirrored the issues revealed at NECC, including selling drugs without a valid prescription, manufacturing large quantities of drugs, and preparing sterile drugs in facilities that were visibly dirty leading to contaminated drug products.

Regulation of Compounding Pharmacies-Federal Role

The FDA regards traditional pharmacy compounding as the combining or altering of ingredients by a licensed pharmacist, in response to a licensed practitioner's prescription for an individual patient, to produce a medication tailored to that patient's special medical needs.[7] In its simplest form, the practice has been used to provide a patient with a medication that is not commercially available, such as adding flavor to a child's dose or removing a dye or preservative for a patient who has specific allergies.

Unlike drug manufacturers, which are required to register with the FDA to make, distribute and sell drugs, there is no requirement for compounding pharmacies to register with the FDA. Furthermore, drug manufacturers are subject to FDA inspections and are required to follow specified safety procedures known as good manufacturing practices (GMPs), as well as report to the FDA any adverse events that are associated with use of their drug products. Compounding pharmacies do not have to follow these same requirements, and like all pharmacies, the oversight of the practice of pharmacy compounding has largely been left to the states.

However, over the last several decades the practice of compounding has expanded and evolved. Today, compounding pharmacies produce injectable medications and other sterile drug products that require the utmost sterility and sophisticated processes and equipment. Internet pharmacies provide mail-order medications to patients in different states whom they will never examine or counsel. Hospitals have moved from in-house to outsourcing pharmacy services for drugs used throughout the hospital. Pharmacies have also expanded their operations to produce thousands of identical drug products, sometimes in advance of prescriptions for easy distribution to medical facilities across the country. In many ways, the activities of these types of pharmacies are more akin to drug manufacturers or distributors, even though the regulatory framework that governs them requires far less regulatory oversight and far less rigorous safety practices.

A Historic View of the Federal Regulation of Compounding Pharmacies

In the early 1990's, after receiving reports of several adverse events associated with compounded medication, the FDA issued a Compliance Policy Guide (CPG) to clarify what constituted appropriate and legitimate pharmacy compounding. The CPG made clear that the agency would use its authority to regulate new drugs in commerce to stop the operation of pharmacies whose activities raise the kinds of concerns normally associated with a drug manufacturer.

In 1997, in order to "clarify the status of pharmacy compounding under federal law,"[8] Congress included provisions to regulate the practice of pharmacy compounding in the Food and Drug Administration Modernization Act (FDAMA).[9] Section 503(A) of the law exempted compounded drugs from the other requirements of the Federal Food, Drug, and Cosmetic Act (FFDCA), most notably the requirements to apply for a new drug approval and follow good manufacturing practices, as long as the pharmacy was licensed in a state, made the drug pursuant to a valid prescription for an individual patient, made the drug using approved ingredients and endorsed standard compounding processes, did not compound inordinate amounts or copies of commercially available drugs, and did not engage in advertising or promotion.[10] Additionally, the exemption from requirements of FFDCA would only hold in states that had entered into a memorandum of understanding (MOU) with the FDA addressing the manner in which the state regulators would investigate and address complaints about compounding pharmacies that were distributing outside the state. In states that did not enter into such an MOU, the exemptions from the provisions of FDAMA only applied if the compounding pharmacy limited its distribution of drugs outside of the state to no more than five percent of its sales.

Before the law took effect, compounding pharmacies challenged the advertising and promotion restrictions of Section 503(A) in federal court.[11] The Ninth Circuit Court found that the Section 503(A) ban on advertising and promotion was an unconstitutional limit on free speech. It also found that the unconstitutional provisions could not be severed from the remainder of Section 503(A), rendering the entire section of law void.[12] Subsequently, the Supreme Court affirmed the lower court rulings that the speech restrictions were unconstitutional, but did not rule on the Ninth Circuit ruling that the unconstitutional provisions could not be severed from the remainder of Section 503(A).

In 2002, FDA issued a new Compliance Policy Guide (CPG), which in large part was very similar to the 1992 CPG and outlined how the agency intended to use its enforcement discretion. FDA stated in its CPG that it would rely heavily on state oversight of compounders and focus its enforcement on compounding pharmacies that were producing large quantities of drugs without valid prescriptions, producing commercially available products, selling drugs wholesale or to third parties for resale, or otherwise violating the new drug, adulteration, or misbranding provisions of the FFDCA.

In 2003, amid rising concerns about compounded drug products, the FDA issued a report entitled "Limited FDA Survey of Compounded Drug Products,"[13] which examined 12 compounding pharmacies that allowed Internet orders. The FDA found that ten of the 29 products it sampled failed one or more standard safety or efficacy tests that were performed on them. A second similar FDA report was issued in 2006.[14] This time the FDA analyzed 36 samples, of which 12 (33%) failed analytical testing using "rigorously defensible testing methodology." The report also mentioned that in the 15 years between 1990 and 2005, "the FDA learned of at least 240 serious illnesses and deaths associated with improperly compounded products."[15]

In 2004, 10 compounding pharmacies brought suit against FDA, challenging more broadly FDA's authority to regulate compounded drugs.[16] The companies charged that compounded drugs were not "new drugs" and were therefore exempt from all of FDA's authority governing new drugs under the FFDCA. In 2008, the Fifth Circuit Court of Appeals found that compounded drugs were in fact "new drugs" and were subject to the FDA's drug approval, adulteration, and misbranding requirements. The Court went further and disagreed with the Ninth Circuit's view on the severability of Section 503(A), effectively reinstating Section 503(A) within the Fifth Circuit's jurisdiction - Texas, Louisiana, and Mississippi.

Throughout the rest of the country not covered by the Fifth or Ninth Circuit decisions, it remains unclear whether the remaining provisions of Section 503(A), that were not ruled unconstitutional, remain in force. FDA has therefore continued to exercise its authority under the FFDCA in accordance with its 2002 Compliance Policy Guide.

However, FDA's 2002 Compliance Policy Guide cannot set binding legal standards. As noted in a recent report by the nonpartisan Congressional Research Service:[17]

> "In contrast to agency rules, which have the force and effect of law, guidance documents are merely considered to be a general statement of

policy. Congress has passed requirements specific to FDA guidance documents, which state that such documents 'shall not create or confer any rights for or on any person, although they present the views of the Secretary on matters under the jurisdiction of the Food and Drug Administration.' Under regulations prescribing FDA good guidance practices, it is stated that 'guidance documents do not establish legally enforceable rights or responsibilities' and 'do not legally bind the public or the FDA.'"

Ambiguity of Agency Authority and Industry Litigation Has Led to Regulatory Confusion

As a result of this ambiguity in law and the non-binding nature of FDA's CPG, when the agency has attempted to utilize its enforcement discretion on a compounding pharmacy, its authority to do so has often been challenged in court. A prime example of this occurred in 2003 when FDA attempted to inspect Wedgewood Village Pharmacy after complaints alleged the pharmacy was making large quantities of drugs without prescriptions.[18] Wedgewood took the agency to court alleging that since they were a licensed compounding pharmacy, FDA had no authority to inspect the facility. Although the court ultimately disagreed with Wedgewood, it took more than a year for the judgment to be issued, and in the meantime the FDA could not collect the evidence necessary to take enforcement action against this pharmacy. On several other occasions when the FDA has been concerned about large quantities of drugs in interstate commerce, it has attempted to inspect the compounding pharmacies in question and has been challenged by the pharmacy and forced to obtain a court-ordered warrant, delaying the agency's ability to address a potential public health crisis.[19]

The FDA has also attempted to rein in the activities of compounding pharmacies through the issuance of dozens of warning letters since 2001 sent to pharmacies believed to have violated federal law governing the production of drugs.[20] While several pharmacies that received these letters ceased problematic activities, some instead challenged FDA's authority to oversee any activities that occur in pharmacies.[21]

The FDA has stated that the multiple court challenges "have produced conflicting case law and amplified the perceived gaps and ambiguity associated with FDA's authority over compounding pharmacies."[22] The agency has called on Congress to take action to prevent another fungal meningitis outbreak and compounded drug tragedy, stating that in the absence of legislation that provides clear authority to the FDA, "this type of incident will happen again, it's a matter of when, not if."

Regulation of Compounding Pharmacies-State Role

State governments, typically through the state boards of pharmacy, have traditionally been the primary entities responsible for all pharmacy practices. It is the duty of the boards to enforce all the laws of the state that pertain to the practice of pharmacy and distribution of drugs. The state boards of pharmacy are also typically the entities responsible for establishing quality assurance and best practices for pharmacies and for examining, licensing, regulating and disciplining pharmacy practitioners. Typically boards are comprised of anywhere from 5-20 members appointed by the state's governor for a specified renewable term. Many boards set aside one or two seats designated for a public member with no pharmacy affiliation who represents the interests of consumers and patients.

Conducting inspections is typically the method that state boards of pharmacy utilize to determine whether pharmacies and pharmacists are in compliance with state regulations for the practice of pharmacy. The methods states use to employ inspections as a compliance tool varies widely. While some states perform routine surprise in-person inspections, other states rely on scheduled announced inspections, while still others primarily rely in whole or in part on pharmacy self-inspections. In a self-inspection, a pharmacy submits responses to a questionnaire on its compliance with laws and regulations. In some states, inspections are driven primarily by the receipt of complaints that warrant an investigation into pharmacy activities.

Historically, any federal regulatory role for the compounding pharmacy sector has been resisted by the industry. In 2007, when draft legislation was circulated to clarify the FDA's role in overseeing the safety of compounding pharmacies, it was immediately denounced by trade associations representing the sector.[23] The trade associations argued that "state boards of pharmacy have done a great job to write compounding standards…There's no way that the FDA will be equipped to handle this."[24] Compounding pharmacy trade organizations have in the past fought vehemently to ensure that pharmacy compounding remained the sole authority of the states, and even issued emails instructing its members what to say to deny FDA authorities access to facilities or samples from facilities that were compounding drugs.[25]

Post-NECC: Massachusetts and Other State Actions

Following the NECC tragedy, Massachusetts Governor Deval Patrick put in place aggressive emergency regulations that required all sterile compounding pharmacies in Massachusetts to immediately and on bi-annual basis report the volume of prescriptions dispensed, states to which prescriptions were distributed and a certification that pharmacies were operating in compliance with all laws and regulations for sterile compounding. The Governor also appointed a special compounding oversight commission that proposed twenty-five additional compounding pharmacy safety measures, many of which were then filed as a legislative package.[26] The legislative package would require a special license for sterile compounding and out-of-state pharmacies, a reorganization of the board members and new authorization for the board to issue fines against pharmacies that violate state laws and regulations. The Governor also immediately hired temporary inspection staff and put in place an enhanced pharmacy inspection schedule that required unannounced inspections of all sterile compounding pharmacies.

Even before the recent fungal meningitis outbreak, in the state of Massachusetts any retail pharmacy that wished to specifically compound sterile injectable drugs[27] had to meet specific regulatory requirements with respect to its clean room facilities and were further expected, but not required, to obtain board of pharmacy approval. In Massachusetts there existed 26 such pharmacies that were known by the state to be producing sterile injectable drugs, including NECC and its sister company Ameridose. However, following the NECC incident, the state requested that pharmacies attest under penalty of law to the scope of its practice. In response to this request, 39 pharmacies,[28] out of a total of 1,179, self-identified as conducting sterile compounding. The Massachusetts Department of Public Health subsequently performed unannounced inspections of these pharmacies to see if they would identify similar issues to those discovered at NECC. These inspections revealed that only 4 were in full compliance with safety standards and required the state to issue 11 immediate full or partial cease and desist orders and 21 deficiency notices to sterile compounding pharmacies.[29]

Since other states have not undertaken such efforts publicly, it is unknown to what extent this rampant disregard for sterile compounding requirements holds true in other states. However, there are several states that have taken steps to increase enforcement activities relating to compounding pharmacies. For example, Iowa has initiated an effort in conjunction with the National Association of Boards of Pharmacy to inspect 600 out-of state pharmacies that ship medications into Iowa. The state of Mississippi adopted an accreditation process in November 2012 for certain sterile compounding pharmacies located in and out of the state.[30]Additionally, Arizona announced plans to establish a task force to determine any regulatory changes that need to be made. The state laws documented in this report are the laws and regulations in place as of November 2012 when states were sent the survey. In the wake of the NECC tragedy, several states, including Maryland, Massachusetts, California, Minnesota, New Hampshire, New Jersey, Oklahoma, South Carolina, Utah, and Virginia have either introduced or passed new legislation or implemented new regulations placing additional restrictions on compounding pharmacies. These post-November 2012 announcements are not reflected in this analysis as most do not represent final enacted and implemented legislation and/or administrative changes.

Additionally, several states, including California, Maryland, Massachusetts, Minnesota, New Hampshire, New Jersey, Oklahoma, South Carolina, Utah, and Virginia have introduced legislation to put stronger controls in place for compounding pharmacies, though to date, none of these legislative solutions have been adopted. It is also expected that additional states will be announcing changes in the coming months.[31]

Among many issues, the NECC tragedy highlighted the inability of states to oversee or manage compounding pharmacies that are located in another state, but that are shipping and selling drugs across state boundaries. In December 2012, the FDA convened a meeting of all 50 state boards of pharmacy to discuss the roles of state and federal entities in the oversight of compounding activities. At this meeting, the states discussed the fact that each state is dependant on other states to regulate pharmacies that are present in their state and this can be a "potential issue, if in fact, there are states that don't have sufficient funding (or) don't have the resources necessary to regulate facilities in their state" because this "can have an impact on every state."[32] A group of state boards present at the meeting expressed their opinion that for "facilities like NECC, there is a role for the FDA to be involved," while pharmacies that are conducting traditional compounding pursuant to a prescription sold only within state borders could be left to most states to adequately oversee. The National Association Boards of Pharmacy (NABP) shared this sentiment, stating that "there's a big disparity among the states in terms of the resources and expertise to regulate compounding" and that "the definition of compounding is that it's patient specific" and once you lose that patient specificity, "It's not a state regulatory issue. It's an FDA issue."[33]

"It's not a state regulatory issue. It's an FDA Issue."
-National Association of Boards of Pharmacy

A previous report issued by Rep. Markey's staff examined media reports, FDA's database of enforcement actions and state board of pharmacy websites and found that state pharmacy boards generally focus enforcement efforts on the types of activities typically associated with traditional pharmacy licensing, such as billing violations, failing to register with the state, failing to have a licensed pharmacist on site, and violations related to the use and distribution of controlled substances.[34] The state boards' of pharmacy efforts are not typically focused on undertaking enforcement actions that relate to the safety or scope of compounding pharmacy practices. Furthermore, state enforcement records are typically not publically available, and when they are available the databases do not allow for keyword searches, preventing the public from easily locating enforcement records or infractions associated with particular pharmacies or medications. When states do keep this information in a public format, the details of the circumstances necessitating the enforcement action is typically lacking, making the information generally useless to a member of the public.

This report summarizes a more extensive investigation into the state oversight of compounding pharmacies, launched to get a better understanding of the information that is maintained internally by state boards of pharmacy, the oversight of pharmacy practices and the actions that states have taken historically to ensure the safety of compounding pharmacies and the products they sell.

INVESTIGATION AND METHODOLOGY

To better understand the role that state boards of pharmacy play in managing the oversight of compounding pharmacies and protecting against another NECC-like tragedy, on November 20, 2012, Reps. Markey, Waxman, Dingell, Pallone and DeGette sent a letter to the boards of pharmacy in all states, territories, and Washington, DC requesting answers to six questions that asked for historical information about compounding pharmacies, pharmacy licensing, inspections of pharmacies, and pharmacy board budgets, as well as information about the structure and policy governing the board (See Appendix C). The questions were designed to examine the degree to which individual states are capable of overseeing the safety of compounding pharmacy practices and enforcing against the type of safety related matters raised by the New England Compounding Center.

With the exception of Rhode Island, Guam, Puerto Rico and the U.S. Virgin Islands, we received a response from all of the state boards of pharmacy queried. When particular responses were incomplete, staff followed up with the board of pharmacy to obtain this information. While a number of states do not have the capacity or system requirements to keep or review historical data, or in some cases simply do not track this information, the states attempted to respond to all questions, providing estimations or projections in areas where hard data was not available or retrievable. Presented below is a summary and analysis of the information provided by the state boards of pharmacy. Additional analysis and findings can be found in Appendix A. Appendix B contains a table that summarizes all state boards of pharmacy responses.

FINDINGS

1. *State boards of pharmacy generally do not know which pharmacies engage in compounding, do not know whether pharmacies ship compounded drugs across state lines, and do not know which pharmacies manufacture large quantities of compounded drugs. In many cases, states are incapable of even providing accurate information regarding the numbers of registered pharmacies in their states.*

To gauge the ability of states to track compounding activities, the state boards were asked to provide information about the number of licensed compounding pharmacies, pharmacies licensed for sterile compounding and pharmacies that sell high volumes of compounded drugs, or sell across state lines.

Key Findings:

- *No state boards require disclosure of volumes of compounded drugs or whether compounded drugs are sold across state lines*
- *Only two states (MO and MS) routinely track the number of compounding pharmacies in their state*
- *Only 32 states were able to provide historical data on the number of licensed pharmacies in their states*

Table 1. States that provided information on the number of compounding pharmacies

State	# licensed in-state pharmacies in most recent year provided	# compoun-ding in-state pharmacies	# in-state sterile compounding pharmacies
AR	755	21*	State does not track
ME	335	3*	State does not track
MO	1,570	442	Did not provide separate number of sterile compounding pharmacies.
MS	1,678 permits issued in the past 10 years	102	State does not track
OR	728	81*	State does not routinely track, but based on inspection records is aware of 11 pharmacies that do sterile compounding and also hold non-resident licenses

* Estimated based on disclosure on initial license application and/or annual renewal. Additional pharmacies may compound without notice.

The vast majority of states allow any pharmacy to compound without a specific compounding license or permit. Forty-seven states and the District of Columbia were unable to provide an exact number of pharmacies that are authorized to compound in the state (See Table 1). Only Missouri and Mississippi require a license or permit for basic drug compounding.

In Mississippi, over the last decade, 102 pharmacies have received specific permits designating them as compounding pharmacies by the Mississippi board of pharmacy. This number does not, however, take into account pharmacies that are no longer in operation, or that simply choose to no longer offer compounding services. Furthermore, because Mississippi does not issue specific permits for facilities engaged in sterile compounding, it's impossible to discern how many of the 102 compounding pharmacies may also be compounding complex and risky sterile drug products.

Missouri's board of pharmacy employs a fairly sophisticated licensing system, which provides different permits for different classes of licensed pharmacies (i.e. internet, veterinary, renal dialysis, etc). In Missouri, all 1,570 pharmacies authorized to dispense medication may perform traditional pharmacy compounding without additional licensure, however the state requires a specific non-sterile compounding permit for pharmacies performing batch compounding from bulk ingredients. Additionally, a sterile compounding permit is required for facilities wishing to engage in sterile compounding activities. In the state of Missouri, 442 pharmacies, approximately one-third of all in-state pharmacies, hold specific permits for compounding both non-sterile drugs from bulk ingredients and for sterile drugs—however, any pharmacy could perform simple compounding services without notifying the state.

None of the states indicated that they track whether pharmacies are selling compounded drugs across state lines, or the volume of compounded drugs that are being produced by a facility.

While Mississippi and Missouri were the only states that indicated the requirement for a specific permit or license to track certain compounding activities, three additional states (Arkansas, Maine, and Oregon) ask that pharmacies indicate on their initial license application whether they intend to engage in compounding activities (See Table 1). Using this information, Maine was able to identify 3 pharmacies and Arkansas 21 pharmacies that provided such an indication. In addition, Oregon, which asks pharmacies to indicate whether they engage in compounding both on the initial license application and on annual renewals, identified 81 compounding pharmacies. It is important to note, however, that in all of these states any pharmacy is authorized to engage in compounding regardless of whether it indicated this intent on their license applications. One additional state, Minnesota, indicated that recent regulatory changes require pharmacies to notify the board and receive approval before they engage in compounding sterile or non-sterile products of any type. However, the state did not indicate how many such pharmacies have self-identified as compounders since this change in state law was implemented.

Several states noted that pharmacies engaging in high-volumes of compounding would trigger the requirement for a manufacturer's registration, as this would be beyond the scope of a state-licensed compounding pharmacy and would instead fall under FDA's jurisdiction. However, it is unclear how any state would know whether this requirement was triggered, as no state indicated that they routinely request information from pharmacies about quantities of compounded drugs or whether such drugs are shipped across state lines as a part of their license or registration.

2. *Only thirteen state boards of pharmacy know which pharmacies are providing sterile compounding services and only five of these states have inspectors that are trained to identify problems with sterile compounding.*

While most states allow any pharmacy to compound without requiring a specific compounding permit or license, a few states have specific requirements for pharmacies that intend to compound sterile drugs. Sterile drugs are more difficult to produce than other drugs and pose significant health threats if improperly produced.

***Key Findings**:*

- *37 state boards of pharmacy (74% of respondents) do not routinely track which pharmacies are providing risky sterile compounding services*
- *Only 19 state boards of pharmacy provide their inspectors with special training to identify problems with sterile compounding*

Including Massachusetts, which even prior to NECC had a clean-room approval process in place for pharmacies that were producing sterile injectable drugs, there were a total of only 13 states (26% of respondents) that were able to provide some estimate as to the number of currently operating sterile compounding pharmacies. Only 5 out these 13 states (ID, NJ, NV,OK and TX) specifically track all pharmacies that perform sterile compounding activities (See Table 2).

- Oklahoma requires in-state retail pharmacies that perform sterile compounding to receive a specific permit and currently has 88 pharmacies permitted to provide these services.
- Alabama requires pharmacies that are compounding preparations for parenteral or intravenous (IV) administration to have a separate certification. Currently there are 103 institutional hospitals and 67 chain and community pharmacies that have parenteral certification. This does not include pharmacies that compound other sterile products whose administration is not through injection.
- Connecticut currently has records of 17 sterile pharmacies, but this doesn't include any hospital pharmacies, which are assumed to be undertaking some sterile compounding.

- As of March 2012, Idaho requires any pharmacy that is engaging in sterile drug preparation to obtain a registration for aseptic environmental control devices, such as laminar flow hoods. Currently, Idaho has provided 77 such registrations, each of which requires an onsite inspection by the board of pharmacy prior to producing sterile drugs.
- While New Jersey doesn't require specific permits or licenses for compounding activities, it does have regulations that require pharmacies to notify the board in advance of compounding sterile preparations. Once notified, the New Jersey board inspects the facility to ensure compliance with regulations related to sterile compounding and then grants approval for the pharmacy to engage in such operations. New Jersey estimated that they have 41 sterile compounding pharmacies.
- California has the second highest number of registered in-state pharmacies (6,761 pharmacies), 266 of which are board-licensed for sterile compounding. However, this number is likely to be a significant underestimate, as pharmacies that are accredited by outside agencies are allowed to perform sterile compounding without additional licensure with the California board of pharmacy.
- The remainder of the 37 states that responded do not systematically track which pharmacies are providing sterile compounding services. The lack of information makes it impossible for pharmacy boards to target enforcement and inspection efforts towards the highest-risk pharmacy operations and appropriately oversee their safe operation.

A previous analysis of state inspection and enforcement records issued by Rep. Edward J. Markey[35] found that state boards of pharmacy do not, as a general rule, undertake enforcement actions that relate to the safety or scope of compounding pharmacy practices. Instead the boards tend to focus efforts on compliance with traditional pharmacy licensing and use of controlled substances. As a result most enforcement actions taken by boards appear to deal with issues such as whether technicians have completed appropriate training hours, appropriate intern supervision by licensed pharmacists, valid and updated registration and licensing documentation and valid distribution of controlled substances.

The survey results indicated that only 19 states (38% of respondents) provide some or all of their inspectors with special training in sterile compounding activities and regulations (See Table 3). An additional 9 states

(CA, IA, IL, KS, MN, MT, NE, UT, and WY) indicated that while there is no special inspector training in compounding or sterile operations, some or all of the inspectors employed by the state are licensed pharmacists and therefore are expected to have basic knowledge of sterile and non-sterile compounding operations.

A few states indicated that after learning about the NECC tragedy, the boards have or are in the process of reevaluating their inspection procedures and developing training specific to sterile compounding for inspectors. For example, Alabama indicated that it is contracting with a consultant to establish particular inspection protocols and training. Additionally, Nebraska and Virginia indicated that it has required inspectors to complete online training in compounding and will provide additional on site training. Mississippi also noted that while state inspectors have limited training in inspecting sterile compounding procedures, they will perform in depth inspections with assistance of inspectors from FDA or the Drug Enforcement Administration (DEA).

3. *States typically do not maintain pharmacy inspection records that enable them to identify systemic or repeated compounding pharmacy safety problems that originate either in-state or out-of-state.*

States were asked to indicate whether they have over the last decade, during the course of an inspection or other oversight activity, identified the kinds of problems that were found at NECC, namely issues with contamination, cleanliness, drug potency, drug safety, bulk manufacturing or other similar concerns. Many states (22 states or 44% of respondents) either do not keep historical records of inspections, or do not track problems relating to compounding. Other states used a combination of inspection records, public complaints and "staff recollections" and were able to indicate a sample of noted problems or disciplinary actions taken against in-state compounding pharmacies due to safety related problems similar to those identified at NECC.

Key Finding:

- *22 state boards of pharmacies do not keep any historical inspection records for compounding pharmacies*

Table 2. Only thirteen states are able to determine the number of pharmacies that perform sterile compounding

State	# licensed in-state pharmacies in most recent year provided	# in-state sterile compounding pharmacies
AL	State does not track	103 institutional hospitals and 67 chain and community pharmacies with parenteral certification
CA	6,761	266 board licensed sterile pharmacies*
CT	670	17 sterile compounding retail pharmacies, not including hospitals
DE	157	State does not track but knows of one sterile facility other than hospitals
IA	1,510	51** sterile compounding retail pharmacies, not including hospitals
ID	445	77
MA	1,179	39^
NJ	2,835	41
NV	709	22**
OK	910	88
TX	6,509	652
VA	1,762	159**
WA	1,425	State does not license, but based on inspections there are 19 pharmacies focusing on parenteral products

* Additional accredited sterile compounders exist, but do not require licensure with the board.

** Any pharmacy could provide sterile compounding services; sterile compounding pharmacies are estimated based on inspection or other information.

^ Represents all sterile compounding pharmacies that self-identified following the fungal meningitis outbreak. Only 4 were found to be in full compliance of safety standards following unannounced inspections.

States reported a total of 2,682 disciplinary actions taken or concerns raised against compounding pharmacies in all 49 states and DC over the last decade. These 2,682 compounding concerns or actions pertained to issues of unsafe storage, compounding copies of commercially available drugs, compounding without a prescription, issues with potency, problems with sterility, use of improper ingredients, and manufacturing large quantities of drugs outside the scope of a pharmacy license It is unclear how many of these compounding-related concerns were evaluated by the boards or rose to the level of formal disciplinary action, as states do not routinely track this information.

Furthermore, because many states do not keep inspection histories, have limited access to details of inspection records, keep them only in an unsearchable paper format, or for a limited timeframe the number of compounding issues reported by the states is likely severely underestimated. These record-keeping practices would make it difficult to determine whether there are particular pharmacies that have a history of violating the law.

As previously indicated, enforcement actions taken against compounding pharmacies are not always publically available on state websites, and when available do not always contain sufficient information for the public to understand the nature of the violation in question, making it impossible for consumers to determine whether a particular facility has had prior safety issues with compounding drugs.[36] When comparing the actions that are disclosed on state websites, and that were analyzed in a previous report,[37] to those disclosed by the states in response to this survey, none of the states reported in their response all of the actions that were listed on their own websites or in media reports.

This calls into question the completeness of the states' online databases and their internal record-keeping systems.

4. *States are unable to effectively police compounding pharmacy activities in other states. Moreover, when issues arise with out-of-state pharmacies, states do not consistently inform the origination state or the FDA.*

Problems with compounded drugs originating from out-of-state pharmacies are common. Nineteen states (38% of respondents) identified 224 issues arising from compounding pharmacies located in other states in the past 10 years. These issues ranged from consumer complaints about potency or safety to discoveries that pharmacies were selling copies of commercially available drugs or distributing samples to medical facilities within the state. In many cases, pharmacies located out-of-state were reprimanded for selling drugs in a different state without the appropriate license or registration. Since three states do not have a requirement for out-of-state pharmacies to be licensed to sell drugs within their state, these states would have no way of even knowing that these sales were occurring.[38]

Many state boards indicated frustration with the lack of control they have over pharmacies located in another state and the absence of a formal mechanism for boards to know about safety related or other issues that arise with out-of- state pharmacies. Please see Table 4, but what follows are highlights from these findings:

> ***Key Finding:***
>
> - *There is no formal mechanism for state boards to know about issues with out-of-state pharmacies*

- The majority (144 out of the 223 or 65 %) of the concerns about out-of-state pharmacies were reported by Iowa, of which just 20 rose to the level of formal disciplinary action.
- North Carolina reported 37 concerns with out-of-state pharmacies. These North Carolina concerns included drugs being linked to several meningitis cases, compounding pharmacies sending samples of drugs directly to medical professionals, pharmacies compounding commercially available drugs and drugs that are not approved for human use and out-of-state pharmacies dispensing drugs in the state of North Carolina without a non-resident pharmacy permit.
- The state of Missouri, which has taken 3 public disciplinary actions against known compounding pharmacies in the last decade issued a cease and desist order to NECC in 2002, but did not indicate the reason for this disciplinary action.
- The state of Colorado also issued a cease and desist order to NECC in 2011, when it was discovered that NECC was manufacturing drugs and shipping them to a hospital pharmacy in violation of Colorado law. At the time, this action was reported to the National Association of Boards of Pharmacy (NABP), the Drug Enforcement Agency (DEA) and FDA, but apparently not directly to the Massachusetts board of pharmacy. When in 2012 the same problem was discovered, Colorado's board of pharmacy notified both the FDA and the Massachusetts board of pharmacy. This second notification occurred approximately two months before the first patient who contracted meningitis from NECC was discovered.

Table 3. States with inspectors trained in sterile compounding

State	# Inspectors Trained in Sterile Compounding Operations in Most Recent Year Reported
AR	3 inspectors and 2 directors involved in investigations and inspections
CO	3
FL	18
GA	14
IN	6
KY	5
LA	5
MD	6.5 full-time pharmacists and 4 technicians split between 2 agencies
MI	5
MO	8
NC	9
ND	2
NJ	7 + 2 part-time inspectors
NM	5
NV	5*
OK	5
TX	7
VA	4
WV	5

* Only a portion of the inspectors receive training.

The remaining 31 responsive states indicated that prior to NECC, they were never made aware of a problem with an out-of-state pharmacy or simply did not track this type of information. Even in cases where there was an identified problem related to an out-of- state pharmacy the direct communication between state boards is limited, as most states never notify other state boards about problems discovered within their borders, even if the pharmacy in question is located in another state. As a result, a state may discover a serious problem with the drugs produced by a pharmacy located in another state, take an action to stop that pharmacy from shipping drugs into its state, but never notify the home state or any other state about the drug safety problem identified. Instead states typically report issues with pharmacies to NABP, and typically this reporting only occurs when the problem is investigated and rises to the level of a formal public disciplinary action. A few states mentioned that they do not report to NABP because they do not share their enforcement activity with non-government entities. Therefore, even NABP would not be a comprehensive source for problems identified at compounding pharmacies.

States also tend to report formal disciplinary actions against pharmacies to the National Practitioner Data Bank (NPDB) and the Healthcare Integrity and Protection Data Bank (HIPDB), the two federal data banks that have been created to serve as repositories of information about health care providers in the United States. Federal law requires that adverse actions taken against a health care professional's license be reported to these data banks. Once reported to these various entities, however, there does not appear to be any systematic manner in which other states are notified of these issues so that they can take proactive action to protect state residents from being harmed. Moreover, these data banks only make individual reports available to the state boards upon request and for a fee.[39] Based on survey results, the only time states seem to routinely inform the FDA about problems with out-of-state pharmacies is when the pharmacy has produced a drug which has caused the death of one or more people.

Table 4. States with information about compounding concerns originating out-of-state in the last decade

State	# Out-of-state Compounding Concerns or Actions	Nature of Concern	Who was Informed
OR	1	Texas Pharmacy sent drugs that were 10 times as potent as indicated on the label, resulting in 3 deaths. The pharmacy was not licensed in Oregon as an out-of-state pharmacy	This particular out-of- state case was investigated by the FDA and Texas BOP
NH	1	Accepted a voluntary license surrender from Infusion Resource located in Massachusetts, based on action taken by FDA and Massachusetts BOP on November 16, 2012	No one
NJ	1	An out-of-state pharmacy notified the NJ Board that they were subject to a disciplinary action as a result of a compounding error in the pharmacy's home state	No one
ID	1	Mail order of invalid prescription drug orders	State of Utah where pharmacy was located

State	# Out-of-state Compounding Concerns or Actions	Nature of Concern	Who was Informed
OK	1*	Pharmacy violated state rules by sending sterile drugs to physician's office for patient pick up	NABP
KS	1	Pharmacy operating without out-of-state registration	The state where the pharmacy was located
GA	1	Selling drugs to an unlicensed facility without prescriptions	Alabama, where pharmacy was located
NY	1**	Pharmacy dispensed contaminated products. Under investigation so details not provided	Unknown
FL	1	Not indicated	No one
NV	2	Compounding a controlled substance without proper in- state registration and selling adulterated drugs	1 case, the State of pharmacy origin was notified. The other case was under CDC and FDA investigation
MN	3	2 unlicensed out-of-state pharmacies and another where an unregistered pharmacy was selling whole sale	The states of origin
LA	3	1 expired in-state permit, 2 pharmacies with histories of adverse events in other states.	NABP as notified in 2 instances
TN	3	1 pharmacy was investigated, but revealed no violation and was dismissed. 2 pharmacies were engaging in manufacturing or wholesaling without licensure	HIPDB and NPDB
MO	3	Not indicated	NABP, NPDB and HIPDB. The board may also provide FDA or state notification when appropriate
WY	6	1 pharmacy was compounding a commercially available drug. 2 pharmacies were operating without a WY license. 1 pharmacy didn't label compounded product for delivery in cold weather. 1 pharmacy had a compounding error. 1 pharmacy was selling a compounded product to a pharmacy for resale.	In 2 of these cases FDA was made aware. In the other cases no was notified

Table 4. (Continued)

State	# Out-of-state Compounding Concerns or Actions	Nature of Concern	Who was Informed
CO	5	1 pharmacy was shipping into the state without a nonresident pharmacy license. 2 pharmacies (one of which was NECC, which was cited twice) were selling manufactured drugs to a hospital pharmacy.	In each case different entities were notified (NABP, HIPDB, stateof origin, FDA and DEA)
CA	9	Only one was investigated based on information from another agency. This one example involved Franck's pharmacy which was dispensing contaminated drugs.	N/A
NC	37*	25 pharmacies were shipping without an in-state license, 1 pharmacy had compo-unded drugs leading to several cases of meningitis, 1 pharmacy was issuing samples, 2 pharmacies were providing drugs for resale by pharmacies, 2 pharmacies were improperly distributing hormone products, 1 pharmacy was dispensing without prescriptions, 3 pharmacies improperly compounded drugs, 1 pharmacy mislabeled drugs, 1 pharmacy compounded a copy of a commercially available drug	In some cases the home state where the pharmacy was located was notified
IA	144	Nature of the issues was not provided, 20 disciplinary actions were taken and several are still pending.	disciplinary actions are typically reported to NABP and HIPDB

* Information was provided for a more limited 6 year period.

** State did not provide historic information, just one pending action.

5. *Despite general increases in state board of pharmacy budgets, the number of pharmacy inspectors has remained consistently low. Furthermore, states usually do not distinguish between inspections of traditional and compounding pharmacies.*

State boards of pharmacy were queried on their historical operating budgets. Forty-six percent of the respondents (23 out of 50 boards) were either unable to provide any information on budgets or were only able to provide data for a more limited time frame than the ten year interval that was requested, citing lack of easy access to historical information.

Nine of these states[40] (18% of respondents) were unable to provide any indication of their operating budget, because the board of pharmacy does not manage its own budget; rather it falls under an umbrella agency that consolidates the budget for all medical and professional boards and manages the expenditures for each board in the state.

Even when factoring in the increase in licensed pharmacies over the last decade, budgets for state boards have generally increased. However, the budgetary constraints vary widely between states.

For example, Nevada currently has a pharmacy board operating budget that provides for approximately $3260 per pharmacy for which the board is responsible for inspecting, while Indiana's board of pharmacy has an operating budget that provides for approximately $173 per pharmacy.

Key Findings:

- *On average states employ just 5 inspectors with responsibility to inspect all pharmacies.*
- *Only 5 state boards of pharmacy were able to provide an estimation of the number of inspections occurring at compounding pharmacies.*

While state pharmacy budgets have modestly grown over the last decade, several recent comments made by state boards of pharmacy representatives[41] and industry indicated that wide- spread budgetary constraints limit board oversight activities.[42]

The difficulty of many states to access data needed to respond to this survey in a timely manner and the inability of many states to track compounding activities in their state does call into question whether state budgets are allowing for the type of oversight that is necessary to ensure the safety of these drugs and the protection of public health.

Even with increasing numbers of pharmacies and general increases in board of pharmacy budgets, on average states employ just 5 inspectors per state, with the most inspectors being utilized in California-30, Ohio-22, Florida-18 and Georgia-14 and the fewest being utilized in Alaska-1,

Vermont-1, Hawaii-1, and Wyoming-1.5. These inspectors are responsible for inspecting all pharmacy activities, and based on survey results, the average number of 5 inspectors has stayed fairly consistent over the last decade. In several cases these inspectors are also split between the board of pharmacy and other professional licensed boards and are responsible for inspecting and investigating all such facilities (for example, dental and medical facilities or other drug distribution facilities including wholesalers) that fall under their purview.

For example, the state of New York indicated that it has 58 inspectors that are responsible for oversight of 50 different professions ranging from elementary educational institutions to pharmacy practice.

Many states increasingly rely on pharmacies to conduct and submit self-inspections. While some states have policies that require all pharmacies to be inspected once a year or every other year, other states only inspect when they receive a complaint about a particular pharmacy or when a pharmacy first receives its license or permit.

Table 5. States that provided data on the number of inspected compounding pharmacies

State	# full-time inspectors	# licensed pharmacies in most recent year provided	#inspections in most recent year provided	# compounding inspections in most recent year provided
CA	30	6,761	2,248	231
MD	10.5	1,819	1,676	106
NJ	7 + 2 part time	2,835	1,026	38
NM	5	283	158	7**
NV	5	709	587	22

* 6.5 pharmacists and 4 technicians split between the Maryland Board of Pharmacy and Division of Drug Control

** indicated that compounding areas within retail pharmacies would be evaluated as a part of routine inspections.

States do not typically track inspections of compounding pharmacies, implying that inspections of entities that engage in riskier compounding behavior such as NECC would not be more likely to be inspected than a traditional pharmacy.

There were only 5 states (California, Maryland, Nevada, New Jersey and New Mexico) that were able to provide some indication of the number of inspections that occurred at compounding pharmacies (a total of 404 compounding pharmacies or 7 percent of the total number of inspections) (See Table 5).

However, in all of these states any pharmacy can engage in compounding, so the data provided are not considered to be fully inclusive of compounding activities.

Since the problems were identified at NECC, Massachusetts enhanced the frequency of its inspection schedule, conducting unannounced inspections of all 37 sterile compounding pharmacies[43] in the state between November 2012 and the end of January 2013.

APPENDIX A

Additional Findings

1. *Over the last decade, forty-seven state boards of pharmacy have seen an increase in the number of licensed pharmacies, but the lack of historical records makes it difficult for most states to accurately track the sector's growth.*

As a part of the survey, the state boards of pharmacy were asked to provide historical data on the number of licensed pharmacies each year for the last decade.

Because of limitations in the way several states maintain records or collect data, there were only 32 states (64% of states) that could provide this information in full. Five states were able to pull responsive records that dated back a more limited timeframe (4-9 years). The remaining 13 states (26% of states) either did not maintain information on the number of licensed pharmacies or could not access this information without a significant time and resource commitment that the state was unable to expend.

According to the survey results, the number of pharmacies in any given state typically increased over the last decade, with the greatest increase happening in the state of Florida where 2,144 new pharmacies opened over the last decade.

Florida was followed in rate of growth by Texas, which opened 1,045 new pharmacies. Florida is also the state which currently has the most licensed operating pharmacies (8,868 pharmacies) followed by California (6,761 pharmacies) and Texas (6,509 pharmacies). In contrast, the states with the fewest licensed pharmacies were New Mexico (34 pharmacies), Vermont (146 pharmacies) and Washington, DC (150 pharmacies). There existed only 3 states (Nebraska, Louisiana, and Oregon) where the number of licensed pharmacies decreased over the last decade, with Oregon losing the most pharmacy operations (149 pharmacies), from 877 in 2003 to 728 pharmacies in 2012.

In total there were approximately 68,000 licensed pharmacy facilities disclosed by the states. However, given poor records kept by states, the fact that some states do not differentiate in their records between in-state and out-of-state pharmacies, and the fact that some states track pharmacies alongside other facilities that dispense drugs, such as distributers and wholesalers, it is impossible to know how accurate this figure is of operating retail and hospital pharmacies in the U.S. Typical estimates made by other industry trade associations place the number of community based retail pharmacies in the U.S at about 56,000.[44]

2. *Regional inconsistencies caused by different case law rulings on the applicability of the 1997 federal pharmacy compounding law have had no apparent impact on the state boards of pharmacy or on the oversight of compounding pharmacies in the Fifth or Ninth Circuits.*

In 1997, Congress enacted the FDA Modernization Act of 1997 (FDAMA) that added a section to FDA law, which exempted compounded drugs from various "new drug" requirements, as long as the compounded drugs met a variety of restrictions.

One of the restrictions was that drug providers were prohibited from soliciting or advertising particular compounded drugs. These speech restrictions were challenged on First Amendment grounds and were struck down by the Supreme Court.

Following this decision, there was controversy over the current status of compounded drugs under the FDA law and whether the remaining provisions that set up standards for compounded drugs remained in affect. The two circuits that addressed this issue took different positions. While the Ninth Circuit determined that the section of law that governed compounding was struck down in its entirety, the Fifth Circuit found that the provisions that didn't deal with speech restrictions are still in effect.

Accordingly, these cases have created an interesting scenario of non-uniform federal law throughout the U.S. In the Fifth Circuit, compounded drugs are specifically exempted from having to apply as a new drug, complying with good manufacturing practices, and certain labeling requirements, as long as certain criteria are met; while in the Ninth Circuit, compounded drugs are subject to all of these requirements that apply to manufactured drugs, but the FDA may exercise discretion in taking action against an entity that violates these provisions.

We compared the states that fall within the Fifth circuit Court's jurisdiction (Texas, Louisiana, and Mississippi) to some of the Western states that fall under the Ninth Circuit Court's jurisdiction and found that there were no significant differences in the composition of the boards, the maintenance of historical records, the number or presence of trained inspectors, or the systematic tracking and oversight of compounding pharmacies (See Appendix Table 1).

3. *All state boards of pharmacy have similar structures and methods to deal with potential conflicts of interest, such as when there is a financial relationship between a pharmacy that is up for disciplinary action and a board of pharmacy member.*

The states were surveyed to get a better understanding of the structure of the state boards of pharmacy. The number of board members varies from 5-17 members, with an average of 8 board members that are typically appointed by the governor, either directly or with input provided by nominations, for a renewable term limit.[45]

All boards are composed of licensed pharmacists, with some states requiring that the pharmacists represent various sectors of the industry, such as chain pharmacies or hospital pharmacies, and additionally may require that some board members are pharmacy technicians or other medical professionals. With the exception of Mississippi, every state has a requirement that board members include at least one public member who is intended to represent the consumer perspective. This public member is typically required to have no past or present affiliation with or financial interest in the practice of pharmacy.

In Mississippi and California the boards are comprised of 11 and 13 members, respectively, and include just one more pharmacist than public members, making these the states with the most public representation. In California, two of the public members are appointed by the Senate Rules and Speaker of Assembly, and the remaining members of the board are appointed by the Governor.

Appendix Table 1. Comparison of Responses for States in the Fifth Circuit Court Jurisdiction (TX, LA, MS) to Those in the Ninth Circuit Court Jurisdiction (OR, NV, CA)

State	#Compounding pharmacies in most recent year	# Sterile compounding pharmacies in most recent year	Specific rules for sterile compounding pharmacies	Inspector training in sterile compounding	#Inspectors in most recent year reported	# Licensed in- state pharmacies(# pharmacies inspected)	# Licensed pharmacies increased or decreased over last decade(by this #)	Reported operating budget	# Board members
TX	State does not track	652	Yes	Yes	7	6,509 (2,140)	Increased (1,045)	$3,717,335 (2003) $6,202,645 (2012)	9 (3 public members)
LA	State does not track	State does not track	Yes	Yes	5	1,758 (793)	Decreased (60)	$1,170,252 (2003) $2,645,000 (2012)	17 (1public member)
MS	102	State does not track	No*	No	6	Not indicated (1,559)	Provided only number of 1,638 new permits issued in past decade	$1,191,615 (2004) $2,026,913 (2013)	7 (0 public members)
OR	81**	State does not track	Yes	No	5	728 (728)	Decreased (149)	$2,358,405 (2001-2003) $6,130,811 (2013-2015)	7 (2 public members)
NV	State does not track	22 (based on inspection records)	Yes	Some inspectors	5	766 (587)	Increased (310)	$1,400,000 (2003) $2,500,000 (2012)	7 (1 public member)
CA	State does not track	266	Pharmacies that compound sterile injections require licensure or accreditation from an approved agency	No	30	6,761 (2,248)	Increased (623)	$7,390,000 (2003) $14,200,000 (2012)	13 (6public members)

* Post-NECC the board adopted a new policy requiring accreditation of sterile compounders.

** Any pharmacy is permitted to compound, numbers estimated based on information provided in application and renewal forms.

Additionally, three states (Arkansas, Florida and Tennessee) require that a member of the board be elderly, usually defined as over the age of 60, and in Arkansas and Tennessee, the laws require that board members include one racial minority member. In Arkansas, this racial minority member must be a licensed pharmacist.

With the exception of Alabama, all states have some policy to deal with conflicts of interest that arise within the board. Typically this policy involves recusal that is either voted on by the board, decided upon by the state Attorney General or other legal counsel, or occurs automatically upon member disclosing an actual or perceived conflict.

Several states also require the board members to provide a financial disclosure that is evaluated by legal counsel to proactively identify conflicts of interest. In many cases the financial disclosure and perceived conflict policies extend to a board member's spouse and immediate family.

While many of the conflict of interest policies require forthrightness of the board members, Washington State has a unique method to proactively avoid conflicts of interest impacting board member decisions. In Washington, any complaints that are handled in a prosecution case are de-identified and only one member of the board knows the identity of the entity that is being prosecuted, leaving board members to decide based purely on the merits and facts of the case.

Additionally, any potential conflicts are screened by the Attorney General's office and state executive ethics board and until a decision is made about the conflict, the member in question is barred from participation in that matter.

Additionally, in Utah, conflicts are proactively investigated prior to appointment to the board, and if any conflicts arise during service, they constitute grounds for removal from the board.

APPENDIX B

Table Summarizing the Responses Received from All State Boards of Pharmacy

State	# Licensed in-state pharmacies in most recent year*estimation	# Compounding in-state pharmacies*additional pharmacies can compound	*Sterile compounding*: Pharmacy registration for sterile injectables—Inspector training—Ability to readily search historical inspection records for problems	# Full-time inspectors	# Pharmacy inspections in most recent year reported(if tracked—# inspections of compounding pharmacies) *estimation ^ includes facilities other than pharmacies	Total # concerns related to in-state compounding pharmacies in time period reported *only a sample reported, not representative of all concerns	# Disciplinary actions for all in-state pharmacies for any reason in the last decade *limited time frame	# Concerns with out-of- state compounding pharmacies	Budget change from first reported year to most recent reported year *provided data for less than decade requested	# non-pharmacist public board members/ total # board members
AK	128	State does not track	No-No-No	1	24	State does not track	101	State does not track	Increased	2/7
AL	1,896*	State does not track	Yes-No-No	6	1,889	1*	102 (includes out of state pharmacies)	State does not track	data not provided	0/5
AR	755	21*	No-Yes-No	5	667	14*	250	Unaware of any	Increased*	2/8
AZ	1,681	State does not track	No-No-Yes	4	894	26	644	Unaware of any	Increased	2/9
CA	6,761	State does not track	Yes-No-Limited	30	2,248 (231)	2,357	82*	9	Increased	6/13
CO	959	State does not track	No-Yes-No	3	1,415^	13*	163	5	Increased	2/7

STATE	# Licensed in-state pharmacies in most recent year*estimation	# Compounding in-state pharmacies*additional pharmacies can compound	*Sterile compounding*: Pharmacy registration for sterile injectables—Inspector training—Ability to readily search historical inspection records for problems	# Full-time inspectors	# Pharmacy inspections in most recent year reported(if tracked—# inspections of compounding pharmacies) *estimation ^ includes facilities other than pharmacies	Total # concerns related to in-state compounding pharmacies in time period reported *only a sample reported, not representative of all concerns	# Disciplinary actions for all in-state pharmacies for any reason in the last decade *limited time frame	# Concerns with out-of- state compounding pharmacies	Budget change from first reported year to most recent reported year *provided data for less than decade requested	# non-pharmacist public board members/ total # board members
CT	670	State does not track	No-No-No	7	84	State does not track	State does not track	Unaware of any	data not provided	2/6
DC	150	State does not track	No-No-Yes	6	200^	64	498*	Unaware of any	Increased	2/7
DE	157	State does not track	No-No-Yes	2	88	State does not track	2	Unaware of any	data not provided	3/9
FL	8,868	State does not track	No-Yes-No	18	4,908	State does not track	324	1	Increased	2/9
GA	2,750	State does not track	No-Yes-No	14	2,500	9*	464	1	data not provided	1/8
HI	2,918	State does not track	No-No-No	1	State does not track	State does not track	State does not track	Unaware of any	data not provided	2/7
IA	1,510	State does not track	No-No-Yes	7	212	36*	611	144	Increased*	2/7
ID	445	State does not track	Yes-No-No	3	360	2	239	1	Increased	1/5
IL	3,299	State does not track	No-No-No	3-6	300*	State does not track	158	State does not track	Increased*	2/9

(Continued)

STATE	# Licensed in-state pharmacies in most recent year*estimation	# Compounding in-state pharmacies*additional pharmacies can compound	*Sterile compounding*: Pharmacy registration for sterile injectables—Inspector training —Ability to readily search historical inspection records for problems	# Full-time inspectors	# Pharmacy inspections in most recent year reported(if tracked—# inspections of compounding pharmacies) *estimation ^ includes facilities other than pharmacies	Total # concerns related to in-state compounding pharmacies in time period reported *only a sample reported, not representative of all concerns	# Disciplinary actions for all in-state pharmacies for any reason in the last decade *limited time frame	# Concerns with out-of- state compounding pharmacies	Budget change from first reported year to most recent reported year *provided data for less than decade requested	# non-pharmacist public board members/ total # board members
IN	1,413	State does not track	No-Yes-Limited	6	685	State does not track	26*	State does not track	Decreased	1/7
KS	893	State does not track	No-No-No	4	864	State does not track	180*	1	Increased*	1/7
KY	2,143	State does not track	No-Yes-Yes	5	2,143*	4	473	Unaware of any	Increased	1/6
LA	1,758	State does not track	No-Yes-Yes	5	793	10	311	3	Increased	1/17
MA	1,179	State does not track	Yes-No-Yes	5	121	8	67	Unaware of any	Increased	2/11
MD	1,819	State does not track	No-Yes-Limited	10.5 split between 2 agencies	1,676 (106)	8	334*	Unaware of any	Increased	2/12
ME	335	3*	No-No-No	2	154	0	732	Unaware of any	Increased	2/7
MI	3,257	State does not track	No-Yes-No	5	1,379	State does not track	100*	State does not track	data not provided	5/11

STATE	# Licensed in-state pharmacies in most recent year*estimation	# Compounding in-state pharmacies*additional pharmacies can compound	*Sterile compounding*: Pharmacy registration for sterile injectables—Inspector training—Ability to readily search historical inspection records for problems	# Full-time inspectors	# Pharmacy inspections in most recent year reported(if tracked—# inspections of compounding pharmacies) *estimation ^ includes facilities other than pharmacies	Total # concerns related to in-state compounding pharmacies in time period reported *only a sample reported, not representative of all concerns	# Disciplinary actions for all in-state pharmacies for any reason in the last decade *limited time frame	# Concerns with out-of- state compounding pharmacies	Budget change from first reported year to most recent reported year *provided data for less than decade requested	# non-pharmacist public board members/ total # board members
MN	1,278	State does not track	No-No-Limited	6	352	11	State does not track	3	Increased	2/7
MO	1,570	442	Yes-Yes-No	8	1,242^	State does not track	137*	3	Decreased	1/7
MS	Current year not provided, 1,678 new permits issued in the past 10 years	102	No-No-Yes	6	1,559	5	430	Unaware of any	Increased	0/7
MT	370	State does not track	No-No-No	2	326	State does not track	160*	Unaware of any	Increased	2/7
NC	2,740	State does not track	No-Yes-Limited	9	289	33	516*	37	Increased	1/6
ND	249	State does not track	No-Yes-No	2	Not indicated	2	49	Unaware of any	data not provided	1/7
NE	524	State does not track	No-No-No	2	133	State does not track	1	Unaware of any	Increased*	1/5
NH	300*	State does not track	No-No-Limited	2	269	2	138*	1	Increased	1/7

(Continued)

STATE	# Licensed in-state pharmacies in most recent year*estimation	# Compounding in-state pharmacies*additional pharmacies can compound	*Sterile compounding*: Pharmacy registration for sterile injectables—Inspector training—Ability to readily search historical inspection records for problems	# Full-time inspectors	# Pharmacy inspections in most recent year reported(if tracked—# inspections of compounding pharmacies) *estimation ^ includes facilities other than pharmacies	Total # concerns related to in-state compounding pharmacies in time period reported *only a sample reported, not representative of all concerns	# Disciplinary actions for all in-state pharmacies for any reason in the last decade *limited time frame	# Concerns with out-of- state compounding pharmacies	Budget change from first reported year to most recent reported year *provided data for less than decade requested	# non-pharmacist public board members/ total # board members
NJ	2,835	State does not track	No-Yes-Limited	7 full-time and 2 part-time	1,026 (38)	225	2,424*	1	Increased	2/11
NM	283	State does not track	No-Yes-Yes	5	158 (7)	0	8	Unaware of any	Increased	3/9
NV	766	State does not track	No-Partial-Yes	5	587 (22)	34	161	2	Increased	1/7
NY	5,244	State does not track	No-No-No	58 responsible for 50 professions	Not indicated	State does not track	Did not respond	1	data not provided	2/14
OH	2,700*	State does not track	No-No-No	22	1,100*	State does not track	State does not track	State does not track	data not provided	1/9
OK	910 (548suppliers)	State does not track	Yes-Yes-Limited	5	639	5	284*	1	Increased*	1/6
OR	728	81*	No-No-No	5	728	State does not track	314	1	Increased	2/7

STATE	**# Licensed in-state pharmacies in most recent year*estimation**	**# Compounding in-state pharmacies*additional pharmacies can compound**	***Sterile compounding*: Pharmacy registration for sterile injectables—Inspector training—Ability to readily search historical inspection records for problems**	**# Full-time inspectors**	**# Pharmacy inspections in most recent year reported(if tracked—# inspections of compounding pharmacies) *estimation ^ includes facilities other than pharmacies**	**Total # concerns related to in-state compounding pharmacies in time period reported *only a sample reported, not representative of all concerns**	**# Disciplinary actions for all in-state pharmacies for any reason in the last decade *limited time frame**	**# Concerns with out-of- state compounding pharmacies**	**Budget change from first reported year to most recent reported year *provided data for less than decade requested**	**# non-pharmacist public board members/ total # board members**
PA	3,200	State does not track	No-No-No	6	1,995	State does not track	400	Unaware of any	Increased*	2/9
SC	2,591	State does not track	No-No-No	4	1,618	State does not track	39*	State does not track	Increased	1/8
SD	295	State does not track	No-No-No	2	Not indicated	0	8*	Unaware of any	data not provided	1/5
TN	2,319 (in-state and out-of-state)	State does not track	No-No-No	5	469	State does not track	262*	3	Increased	1/7
TX	6,509	State does not track	Yes-Yes-Yes	7	2,140	18*	684	Unaware of any	Increased	3/9
UT	1,331	State does not track	No-No-No	4	142	State does not track	365	State does not track	data not provided	1/7
VA	1,754	State does not track	No-Yes-No	4	647	State does not track	412	State does not track	data not provided	2/10
VT	146	State does not track	No-No-No	1	Not indicated	State does not track	Did not respond	Unaware of any	data not provided	2/7
WA	1,425	State does not track	No-No-Limited	11	1,198	36	29	Unaware of any	Increased	2/7

(Continued)

STATE	# Licensed in-state pharmacies in most recent year*estimation	# Compounding in-state pharmacies*additional pharmacies can compound	*Sterile compounding*: Pharmacy registration for sterile injectables—Inspector training—Ability to readily search historical inspection records for problems	# Full-time inspectors	# Pharmacy inspections in most recent year reported(if tracked—# inspections of compounding pharmacies) *estimation ^ includes facilities other than pharmacies	Total # concerns related to in-state compounding pharmacies in time period reported *only a sample reported, not representative of all concerns	# Disciplinary actions for all in-state pharmacies for any reason in the last decade *limited time frame	# Concerns with out-of- state compounding pharmacies	Budget change from first reported year to most recent reported year *provided data for less than decade requested	# non-pharmacist public board members/ total # board members
WI	1,220	State does not track	No-No-No	9	67	State does not track	14	Unaware of any	Increased (estimate)	2/7
WV	636	State does not track	No-Yes-No	5	Not indicated	State does not track	Did not respond	Unaware of any	Increased*	2/7
WY	174	State does not track	No-No-No	1.5	174-200*	15	29	4	Increased*	1/8

APPENDIX C

Letter to State Boards of Pharmacy sent by Reps. Edward J. Markey (D-Mass.), Henry A. Waxman (D-Calif.), John Dingell (D-Mich.), Frank Pallone (D-N.J.) and Diana DeGette (D-Colo.)

FRED UPTON, MICHIGAN
CHAIRMAN

HENRY A. WAXMAN, CALIFORNIA
RANKING MEMBER

ONE HUNDRED TWELFTH CONGRESS

Congress of the United States

House of Representatives

COMMITTEE ON ENERGY AND COMMERCE
2125 RAYBURN HOUSE OFFICE BUILDING
WASHINGTON, DC 20515-6115

Majority (202) 225-2927
Minority (202) 225-3641

November 20, 2012

Mitzi Ellenburg
Interim Director
Alabama State Board of Pharmacy PO Box 381988
Birmingham , AL 35238-1988

Dear Ms. Ellenburg:

The recent deadly outbreak of fungal meningitis linked to spinal injections of a pain-reliving steroid produced by the New England Compounding Center in Framingham, Massachusetts, has raised serious questions about the oversight of compounding pharmacies and the appropriate role for federal and state regulators in this sector.

Traditional compounding pharmacies that produce specialized drug formulations in response to an individual prescription play an important role in our healthcare system. However, we learned at our November 14 hearing that there have been a growing number of large compounding pharmacies engaging in mass production of drugs to be sold in bulk, often over state lines, in a manner much more akin to drug manufacturers. While drug manufacturers are required to meet stringent standards enforced by the Food and Drug Administration (FDA), compounding pharmacies, like all pharmacies, are licensed and overseen primarily by the states, allowing some of these facilities to evade federal safety requirements, putting public health and safety at risk.

With state budgets stretched thin, we are concerned that many state boards of pharmacies lack adequate resources and may have an insufficient number of inspectors trained in overseeing and inspecting large compounding pharmacies that engage in highly complex sterile processes as part of their operation as *de facto* manufacturers. To better understand how state boards of pharmacies manage the oversight of compounding pharmacies, we respectfully request that you respond to the following questions:

1. Please describe the role played by your pharmacy oversight board, the number of members of your pharmacy oversight board, the process for appointing such members and the qualification s required of such members, and the process for managing conflicts of interest on your pharmacy oversight board.
2. For each of the past ten years, please indicate how many pharmacies in your state were a) licensed; b) licensed as compounding pharmacies; c) licensed to conduct sterile compounding; and d) licensed to sell high volumes of compounded drugs or sell compounded drugs across state lines (if your state does not require a specific license for sterile compounding activities, or for high volumes of compounded drugs, or for compounding pharmacies more generally, please also indicate this in your response).
3. Does your state require special licensing, safety requirements or certification for sterile compounding pharmacies or for compounding pharmacies that sell high volumes of compounded drugs or sell compounded drugs across state lines? If so, please provide a copy of or citation to these requirements. Are inspectors with responsibility for inspecting pharmacies in your state trained or certified in such sterile or high volume compounding procedures?
4. For each of the past ten years, please list a) the operating budget of the state board, b) the number of inspectors with responsibility for inspecting pharmacies, c) the number of pharmacy inspections conducted in your state, and d) the number of inspections of compounding pharmacies that took place in your state.
5. For each of the past ten years, please list a) the number of times an inspection or other oversight function of the state board identified concerns (e.g., contamination, cleanliness, drug potency, drug safety, bulk manufacturing or other similar problems) during an inspection of a compounding pharmacy, b) what the nature and outcome of the concern was, including the type of pharmacy, type of concern,

involved medication, and any disciplinary actions taken, c) whether and how the state board informed the FDA, the National Association of Boards of Pharmacy, other state Boards of Pharmacy, or other regulatory authorities regarding the investigation or disciplinary action taken, and d) the total number of disciplinary actions taken against any type of pharmacy for any reason by your board in each year.

6. With the exception of the most recent issues with the New England Compounding Center, in each of the past ten years, how many times has the board identified concerns with the activities of a compounding pharmacy located in another state that shipped drugs into your state? For each such occurrence, please describe the nature and outcome of the concern, including whether your board notified the FDA, the National Association of Boards of Pharmacy, or the board of pharmacy of the state in which the compounding pharmacy was located. Please additionally provide a copy of all documents (including emails, letters, telephone logs or faxes) in possession of your board related to such notification, including any response the board may have received.

Thank you for your assistance and timely response. Please provide your response no later than Friday December 7, 2012. Should you have any questions about this request please contact Dr. Avenel Joseph or Dr. Michal Freedhoff of Rep. Markey's staff at 202- 225-2836.

Sincerely,

Edward J. Markey
Member of Congress

Henry A. Waxman
Ranking Member
Energy and Commerce Committee

John D. Dingell
Chairman Emeritus
Energy and Commerce Committee

Frank Pallone, Jr.
Ranking Member
Subcommittee on Health

Diana DeGette
Ranking Member
Subcommittee on Oversight and Investigations

End Notes

[1] Appendix C includes a copy of this request.

[2] Food and Drug Administration, *Form 483 Issued to Barry Cadden* (Oct. 26, 2012) and Massachusetts Department of Public Health, Board of Registration in Pharmacy Report, *New England Compounding Center: Preliminary Investigation Findings* (Oct. 23, 2012).

[3] The Committee on Energy and Commerce Hearing on "The Fungal Meningitis Outbreak: Could it Have Been Prevented?" Majority Memorandum. November 12, 2012.

[4] Food and Drug Administration, *Form 483 Issued to Barry Cadden* (Oct. 26, 2012)

[5] Colorado Board of Pharmacy also discovered NECC engaging in illegal distribution of compounded drugs to hospitals in April 2011, but did not notify the Massachusetts Board of Pharmacy at this time.

[6] October 29, 2012 Compounding Pharmacies, Compounding Risk. See: http://markey.house.gov/press-release/new-markey-report-reveals-current-outbreak-least-23-deaths-86-serious-illnesses

[7] November 12, 2012. Statement of Dr. Margaret Hamburg, Commissioner of Food and Drugs before the Subcommittee on Oversight and Investigations Committee on Energy and Commerce.

[8] Conference Report H.Rept. 105-399 at 94. November 9, 1997.

[9] Pub.L. No.105-115, (1997)

[10] Pub.L. No.105-115, §503(A) (1997).

[11] *Western States Medical Center, et al. v. Shalala*, 69 F. Supp. 2d 1288 (D. Nev. 1999).

[12] *Thompson v. Western States Medical Center*, 535 U.S. 357 (2002).

[13] http://www.fda.gov/Drugs/GuidanceComplianceRegulatoryInformation/Pharmacy Compounding/ucm155 725.htm

[14] http://www.fda.gov/Drugs/GuidanceComplianceRegulatoryInformation/Pharmacy Compounding/ucm204 237.htm

[15] Ibid.

[16] Med. Ctr. Pharmacy v. Mukasey, 536 F.3d 383 (5th Cir. 2008).

[17] FDA's Authority to Regulate Drug Compounding: A Legal Analysis. Jennifer Staman, October 17, 2012.

[18] In the Matter of Establishment Inspection of Wedgewood Pharmacy, 421 F.3d 263 (3rd Cir. 2005) See: http://www.fda.gov/downloads/iceci/enforcementactions/enforcementstory/enforcementstoryarchive/ucm0 91066.pdf

[19] See for example In the Matter of Establishment Inspection of Wedgewood Pharmacy, 421 F.3d 263 (3rd Cir. 2005).

[20] See FDA's database on warning letters:http://www.fda.gov/ICECI/EnforcementActions/WarningLetters/default.htm

[21] *See*: Warning Letter (NEW-06-07W) from Gail T. Costello, Dist. Dir., New England Dist. Office, FDA, to Barry J. Cadden, Dir. of Pharmacy, New England Compounding Center (Dec. 4, 2006) and Letter from Barry J. Cadden, Dir. of Pharmacy, New England Compounding Center, to Compliance Officer, New England Dist. Office, FDA et al., at 1 (Jan. 5, 2007). *See* also: FDA Acting Director, Steven Galston testimony before the Senate Committee on Health, Education, Labor and Pensions hearing on the "Federal and State Role in Pharmacy Compounding and Reconstitution: Exploring the Right Mix to Protect Patients. October 23, 2003.

[22] Testimony of FDA Commissioner Margaret Hamburg before the House Subcommittee on Oversight and Investigations Committee On Energy and Commerce hearing on "The Fungal Meningitis Outbreak: Could it Have Been Prevented?" November 14, 2012.

[23] See: http://drugtopics.modernmedicine.com/drugtopics/Community+Pharmacy/New-bill-on-pharmacy- compoundingstirs-concern/ArticleStandard/Article/detail/414436

[24] See: http://drugtopics.modernmedicine.com/news/new-bill-pharmacy-compounding-stirs-concern

[25] Walt Bogdanich and Sabrina Tavernise. "U.S. Concern Over Compounders Predates Outbreak of Meningitis." *The New York Times* 23 October 23 2012 A1.

[26] See: http://www.mass.gov/governor/pressoffice/pressreleases/2013/0104-compounding-industry- legislation.html

[27] Note that this represents a subset of all sterile compounders in Massachusetts

[28] Two of these 39 pharmacies attested to either closing their operations or ceasing the practice of sterile compounding, leaving 37 self-identified as conducting sterile compounding. This number does not include NECC, Ameridose or Alaunus.

[29] Kay Lazar and Chelsea Conaboy. "Just 4 of 37 Massachusetts Compounding Pharmacies Passed Surprise Health Inspections" *The Boston Globe* 6 February 2013 online.

[30] In Mississippi, accreditation for sterile pharmacies excludes hospitals preparing sterile injectable drugs for use within 24 hours and pharmacies compounding these drugs for administration to institutional patients within 72 hours. Mississippi also put in place an accreditation requirement for non-sterile compounding pharmacies that compound more than 25 % of their total prescription volume.

[31] This report notes some of these pending state changes, but is not an exhaustive discussion. In many instances the policy positions are not final, and in others, states may have initiated such efforts but not reported them to Congressional requesters.

[32] FDA Framework for Pharmacy Compounding: State and Federal Roles. December 19, 2012.Transcript.

[33] February 1, 2013: NPR Diane Rehm Show: Safety Concerns at Compounding Pharmacies, Statement by Carmen Catizone, Executive Director of the National Association of Boards of Pharmacy.

[34] October 29, 2012 Compounding Pharmacies, Compounding Risk. See: http://markey.house.gov/press- release/new-markey-report-reveals-current-outbreak-least-23-deaths-86-serious-illnesses

[35] Compounding Pharmacies, Compounding Risk; issued October 29, 2012. See: http://markey.house.gov/press-release/new-markey-report-reveals-current-outbreak-least-23-deaths-86-I serious-illnesses

[36] Compounding Pharmacies, Compounding Risk; issued October 29, 2012. See: http://markey.house.gov/press-release/new-markey-report-reveals-current-outbreak-least-23-deaths-86-serious-illnesses

[37] Ibid.

[38] Three states (Massachusetts, Georgia and Pennsylvania) do not require out-of-state pharmacies to hold a license within their state. Following the NECC linked fungal meningitis outbreak, the Governor of Massachusetts filed legislation to make several reforms, including a requirement that all non-resident pharmacies be licensed and subject to the regulations of the board of pharmacy.

[39] Informal communication with Massachusetts Department of Public Health.

[40] States with boards of pharmacy that fall under a consolidated umbrella agency: Connecticut, Delaware, Georgia, Hawaii, Michigan, New York, Utah, Vermont, Virginia. Wisconsin also falls under an umbrella agency, but was able to provide an estimate of its budget.

[41] For example, informal communications between some state board of pharmacy contacts and Rep. Markey's staff indicated that states with consolidated professional boards and those who rely on annual state appropriations for their budgets have a difficult time with securing the resources needed to adequately oversee the full range of licensed entities, while independent boards funded directly via pharmacy licensing fees typically have more budget stability

[42] See: December 19, 2012: FDA Framework for Pharmacy Compounding: State and Federal Roles. Statement by Dr. Cody Wiberg, Minnesota Board of Pharmacy. February 1, 2013:

NPR Diane Rehm Show: Safety Concerns at Compounding Pharmacies, Statement by Dr. David Miller of the International Academy of Compounding Pharmacists.

[43] There were 39 sterile compounding pharmacies in Massachusetts, but 2 of these ceased operating as sterile compounders, leaving 37 subject to inspections.

[44] http://www.iacprx.org/displaycommon.cfm?an=1&subarticlenbr=277

[45] New York Regents Board appoints the Board of Pharmacy members, New Hampshire State Board of Health appoints its Board members.

In: Drug Compounding Pharmacies
Editor: Seth Sharpe
ISBN: 978-1-62808-176-3

Chapter 4

COMPOUNDING PHARMACIES, COMPOUNDING RISK*

Edward J. Markey

EXECUTIVE SUMMARY

The recent outbreak of fungal meningitis caused by contaminated injectable steroids manufactured by New England Compounding Center (NECC) has currently killed 25 people and sickened 344 people in 18 states. It has also raised many questions, starting with a fundamental one: How could this have happened? This report describes the nature of the regulatory oversight and gaps in legal authority that led to one of the worst public health tragedies in recent U.S. history.

Traditional pharmacy compounding occurs when a pharmacist "combines, mixes, or alters various drug ingredients to create a medication for an individual patient in response to a practitioner's prescription."[1] The practice has typically been used to provide a patient with a medication that is not commercially available, such as adding flavor to a child's dose or removing an allergen from a drug for a patient with an allergy.

* This is an edited, reformatted and augmented version of the staff of the office of Congressman, dated October 29, 2012.

In the past few decades, however, the practices associated with pharmacy compounding have evolved. Internet pharmacies provide mail-order medications to patients in different states who they will never examine or counsel. Large compounding pharmacies produce vast quantities of drugs that are copies of commercially available medications even though these drugs do not differ in any meaningful way from the commercial product they are mimicking. And while the Food and Drug Administration (FDA) has clear authority to regulate the development and manufacture of new drugs, the regulation of traditional pharmacy practices has historically been left largely to the states.

In 1997, Congress attempted to clarify the FDA's authority over the evolving activities of compounding pharmacies as part of the FDA Modernization Act (FDAMA). But this law – as well as FDA's efforts to use its existing authority under the Federal Food, Drug and Cosmetic Act to regulate compounding pharmacies - soon became mired in litigation and uncertainty as some compounding pharmacies and their trade associations insisted that FDA should not have any authority over their activities. For example, a 2011 press release[2] about a court's decision in a veterinary compounding case that was issued by the International Academy of Compounding Pharmacists (IACP) stated that the court's decision reaffirmed "what IACP has said for years – the FDA does not have jurisdiction over the traditional practice of pharmacy compounding. That is the sole authority of the state Boards of Pharmacy."

In this report, Congressman Edward J. Markey's (D-Mass.) office examines the respective regulatory roles played by the FDA and the state Boards of Pharmacy by analyzing media accounts and all publicly available safety-related compounding pharmacy enforcement actions taken by the FDA or any of the 50 states, Puerto Rico or the District of Columbia. The conclusions of this analysis indicate that:

- FDA's efforts to assure the safety of compounding pharmacies have been challenged at every juncture by some members of the compounding pharmacy sector.
- Since June 2001, the FDA has attempted to rein in the activities of dozens of pharmacies through the issuance of warning letters, all of which are publicly available.
- FDA records and media accounts indicate problematic compounding pharmacy practices or adverse medical incidents related to such practices occurred in at least 34 states.

- FDA records document 23 deaths and at least 86 serious illnesses or injuries associated with these practices (not including the recent deaths and illnesses that are attributable to fungal meningitis caused by medications produced by New England Compounding Center). These totals should be considered to be conservative, since in many cases the reviewed documents noted the existence of adverse events but did not specify the type or quantity; and in other cases, the warning letters may have been issued prior to a full realization of the health impact of the alleged violation. These documents do not include enforcement actions taken by state regulators or improper practices and adverse events that never led to FDA action.
- Violations noted by the FDA include:
  - numerous instances of compounding pharmacies selling copies of commercially- available drugs;
  - numerous instances of compounding pharmacies selling drugs made using ingredients that were not FDA-approved, or substances that were recalled for safety or effectiveness reasons;
  - numerous serious violations of good manufacturing practices, including cases where facilities purporting to be sterile were visibly dirty, and cases where contamination of the drug product was known to have occurred; and
  - numerous instances of compounding pharmacies whose practices consisted of either selling drug products without a valid prescription or manufacturing large quantities of drug products, such that they were more akin to drug manufacturers than pharmacies.

By contrast, an examination of the efforts of the state Boards of Pharmacy does not demonstrate a consistent role for state regulators in the efforts to assure the safety of drugs made by compounding pharmacies:

- The publicly-available information regarding the enforcement activities of State Boards of Pharmacy typically involve traditional types of violations by individual pharmacies or pharmacists (such as billing violations or failure to have a licensed pharmacist onsite) or the distribution of controlled substances via the falsification of prescription records or other means.
- State enforcement records related to the safety of compounding pharmacy practices were not typically found in the enforcement

records reviewed by Rep. Markey's staff. The exceptions to this general finding were found in only six States (Arizona, California, Missouri, New York, North Carolina and Rhode Island).

- Most state Board websites do not allow for keyword searches, preventing members of the public from easily locating or downloading enforcement records associated with particular pharmacies or infractions associated with 'compounding pharmacies' or particular medications.
- Instead of keyword search capability, state enforcement action records related to pharmacies are often limited to alphabetical or temporal lists or summaries of violations that are not themselves searchable. These lists sometimes do not contain sufficient information to understand the nature of a violation in question. Only four state Board websites (Arizona, California, New Jersey and North Carolina) allowed for full keyword searches and enforcement document downloads.
- In many cases, the information available on the state Board websites is limited. For example, while the Massachusetts state Board (and many others) allows for searches of individual pharmacy licensees, the actual enforcement documents that detail the nature of any complaint and the resolution thereof are only available by request to the Board[3]. Additionally, the Massachusetts state Board (and many others) does not post any records associated with complaints that have been voluntarily resolved or dismissed – so the public would not have been able to learn of the earlier safety problems and 2006 consent decree entered into by New England Compounding Center via a search of the Board website.

This analysis makes clear that state regulators are not, or cannot, perform the same sort of safety-related oversight of compounding pharmacy practices that FDA has historically undertaken. But it is also clear that absent clear new statutory authority, FDA's efforts will ultimately be constrained by gaps in regulatory authority and thwarted by an industry that has historically resisted a federal role for the oversight of its activities.

SAFETY PROBLEMS AT COMPOUNDING PHARMACIES: THE FEDERAL ROLE

Traditional pharmacy compounding occurs when a pharmacist "combines, mixes, or alters various drug ingredients to create a medication for an individual patient in response to a practitioner's prescription.[4]" The practice has typically been used to provide a patient with a medication that is not commercially available, such as adding flavor to a child's dose or removing an allergen from a drug for a patient with an allergy.

For decades, pharmacy compounding practices were regulated solely by state Boards of Pharmacy, and the drugs made by compounding pharmacies were not required by Food and Drug Administration (FDA) to be approved as "new drugs".

In the early 1990s, the FDA became concerned[5] that some compounding pharmacies were exceeding the traditional practices of providing compounded drugs for individual patients with prescriptions as it became aware of large-scale compounding drug pharmacies whose activities seemed more akin to drug manufacturing. In response, in 1992 the FDA issued a Compliance Policy Guide for compounding pharmacies in which it stated that compounding pharmacies were not explicitly exempted from its authority, and that it would exercise its discretion to enforce its authority "when the scope and nature of a pharmacy's activity raises the kind of concerns normally associated with a manufacturer."[6]

In 1997, legislation was enacted as part of the FDA Modernization Act (FDAMA) to clarify FDA's oversight and regulation of compounding pharmacies whose activities exceeded traditional pharmacies' practices.

That legislation exempted compounding pharmacies from certain FDA regulations as long as the pharmacies compounded drugs for individual patients with valid prescriptions for the drugs, as long as the drugs being compounded were made using approved ingredients and endorsed standard manufacturing processes, and as long as the drugs were not copies of a commercially-available drug.

Additionally, the exemption from federal regulation would only hold in states that had entered into a memorandum of understanding with the FDA that addressed the manner in which the state regulators would investigate and address complaints about compounding pharmacies that were distributed outside the state, or, in states that did not enter into such a memorandum of understanding, if the compounding pharmacy did not distribute more than five percent of its drug products outside the state.

The 1997 legislation also prohibited pharmacies from advertising and promoting their compounded drug products. It was this provision that was struck down in 2002 by the U.S. Court of Appeals for the Ninth Circuit on the grounds that it violated pharmacies' Constitutional right to free speech. The court also ruled that the advertising provision was not severable from the rest of the law, and thus invalidated all the provisions enacted in 1997.

The Supreme Court upheld the Ninth Circuit's decision on the advertising provisions in 2002, but did not rule on whether the rest of the provisions must also be struck down. In response the FDA issued a new Compliance Policy Guide[7] which stated that in light of the Court's decision, the FDA would exercise enforcement discretion for compounding pharmacies that depended in part on whether a compounding pharmacy was found to have engaged in any of nine specific activities related to the scale of its operations and the safety of the compounds it was using. For other violations, it would defer to state regulatory authorities.

The nine specific circumstances listed by FDA as having the potential for FDA enforcement were:

1. Compounding of drugs in anticipation of receiving prescriptions, except in very limited quantities in relation to the amounts of drugs compounded after receiving valid prescriptions.
2. Compounding drugs that were withdrawn or removed from the market for safety reasons. Appendix A provides a list of such drugs that will be updated in the future, as appropriate.
3. Compounding finished drugs from bulk active ingredients that are not components of FDA approved drugs without an FDA sanctioned investigational new drug application (IND) in accordance with 21 U.S.C. § 355(i) and 21 CFR 312.
4. Receiving, storing, or using drug substances without first obtaining written assurance from the supplier that each lot of the drug substance has been made in an FDA- registered facility.
5. Receiving, storing, or using drug components not guaranteed or otherwise determined to meet official compendia requirements.
6. Using commercial scale manufacturing or testing equipment for compounding drug products.

7. Compounding drugs for third parties who resell to individual patients or offering compounded drug products at wholesale to other state licensed persons or commercial entities for resale.
8. Compounding drug products that are commercially available in the marketplace or that are essentially copies of commercially available FDA-approved drug products. In certain circumstances, it may be appropriate for a pharmacist to compound a small quantity of a drug that is only slightly different than an FDA-approved drug that is commercially available. In these circumstances, FDA will consider whether there is documentation of the medical need for the particular variation of the compound for the particular patient.
9. Failing to operate in conformance with applicable state law regulating the practice of pharmacy.

In 2003, the FDA issued a report[8] entitled "Limited FDA Survey of Compounded Drug Products" which examined 12 compounding pharmacies that allowed Internet orders. The FDA found that ten of the 29 products it sampled failed one or more standard safety or efficacy tests that were performed on them. A second FDA report[9] in 2006 was issued, this time sampling 198 products. According to this report,

> "125 were active pharmaceutical ingredients (APIs) and 73 were compounded finished drug products. The samples comprised three major drug classes: female hormone products, inhalation products, and local anesthetic products. All 125 API samples passed analysis. Of the 73 compounded finished drug products, sixteen samples were not analyzed because the expiration dates on the samples elapsed before analysis. The remaining 57 samples were analyzed, but the results of the analyses for 21 of these samples were deemed unusable for various reasons and excluded from the survey. Of the remaining 36 samples, 12 (33%) failed analytical testing using rigorously defensible testing methodology." The report also stated that "from 1990 to 2005, FDA learned of at least 240 serious illnesses and deaths associated with improperly compounded products."

In 2008, the Fifth Circuit agreed that the advertising provision in the 1997 legislation infringed on pharmacies' free speech. In contrast to the Ninth Circuit, however, the Fifth Circuit held that this provision was severable from the rest of the law. In response, the FDA issued a statement that the entire 1997 provision on compounding drugs remained invalid in all of the country except for the states of the Fifth Circuit (Texas, Louisiana, and Mississippi) and that the FDA would continue to operate under the "Compliance Regulatory Guide". However, according to a recent Congressional Research Services report:[10]

> "In contrast to agency rules, which have the force and effect of law, guidance documents are merely considered to be a general statement of policy. Congress has passed requirements specific to FDA guidance documents, which state that such documents "shall not create or confer any rights for or on any person, although they present the views of the Secretary on matters under the jurisdiction of the Food and Drug Administration." Under regulations prescribing FDA good guidance practices, it is stated that guidance documents do not establish legally enforceable rights or responsibilities and do not legally bind the public or the FDA."

Historically, trade associations representing compounding pharmacies have resisted all efforts to regulate their activities at the federal level.[11,12] The original lawsuit that led to the 2002 invalidation of the 1997 law to allow the FDA to oversee some compounding pharmacies' activities was filed by eight specialty compounding companies. The suit that led the Fifth Circuit to affirm (in 2008) the FDA's right to enforce its authority to consider compounded drugs "new drugs" subject to FDA authority was brought by ten pharmacies in 2004, which asserted that the FDA had no such authority. And in 2004, FDA was forced to go to court[13] in order to inspect Wedgewood Pharmacy, a compounding pharmacy that was operating in a manner that was more akin to a manufacturer, when Wedgewood argued that it was exempt from FDA inspection authority under the Federal Food, Drug and Cosmetic Act. While the court ultimately disagreed with Wedgewood, the fact that the litigation occurred in the first place underscores the lack of clear regulatory authority for the activities of compounding pharmacies.

Additionally, in 2007 when legislation was again circulated to clarify the FDA role in overseeing the safety of compounding pharmacies, it was immediately denounced[14] by trade associations representing the sector. In fact, nine associations representing pharmacists (the American Pharmacists Association, National Community Pharmacists Association, International Academy of Compounding Pharmacists (IACP), American College of Apothecaries, American Society of Consultant Pharmacists, National Alliance of State Pharmacy Associations, Massachusetts Pharmacists Association, North Carolina Association of Pharmacists, and Kansas Pharmacists Association) urged Senators Kennedy, Roberts and Burr to reconsider their plans to introduce the legislation. According to L.D. King, then executive director of IACP, "State boards of pharmacy have done a great job to write compounding standards...There's no way that the FDA will be equipped to handle this."

In the wake of the multiple deaths caused by the apparent contamination of medications produced by New England Compounding Center, a spokesman for the Health and Human Services Department stated that, "This outbreak began at a compounding pharmacy and the Food and Drug Administration has very limited authority over what these facilities produce. We urge Congress to give FDA the authority it needs to ensure these kinds of outbreaks do not happen again."

Rep. Markey's office examined media reports as well as the FDA's database of warning letters and other materials related to compounding pharmacies. These efforts were not exhaustive and some incidents may have been inadvertently omitted. Appendix A is an attempt to list the enforcement action the FDA has undertaken and the long history of safety issues associated with the practices of some compounding pharmacies across the United States.

In summary:

- Since June 2001, the FDA has attempted to rein in the activities of dozens of pharmacies through the issuance of warning letters, all of which are publicly available.
- Records indicate problematic compounding pharmacy practices or adverse medical incidents related to such practices occurred in at least 34 states, including AL, AR, AZ, CA, CT, DC, DE, FL, ID, IL, IN, KY, LA, MA, MD, MI, MN, MO, MS, NC, ND, NE, NH, NJ, NY, OH, OK, PA, PR, SC, TX, UT, VA, and WY.
- Records document 23 deaths and at least 86 serious illnesses or injuries associated with these practices (not including the recent deaths and illnesses that are attributable to fungal meningitis caused by medications produced by New England Compounding Center). These totals should be considered to be conservative, since in many cases the reviewed documents noted the existence of adverse events but did not specify the type or quantity; and in other cases, the warning letters may have been issued prior to a full realization of the health impact of the alleged violation. These documents do not include enforcement actions taken by state regulators or improper practices and adverse events that never led to FDA action.
- Violations noted by the FDA include:

- numerous instances of compounding pharmacies selling versions of commercially- available drugs;
- numerous instances of compounding pharmacies selling drugs made using ingredients that were not FDA-approved, or substances that were recalled for safety or effectiveness reasons;
- numerous serious violations of good manufacturing practices, including cases where facilities purporting to be sterile were visibly dirty, and cases where contamination of the drug product was known to have occurred; and
- numerous instances of compounding pharmacies whose practices consisted of either selling drug products without a valid prescription or manufacturing large quantities of drug products, such that they were more akin to drug manufacturers than pharmacies.

SAFETY PROBLEMS AT COMPOUNDING PHARMACIES: THE STATE ROLE

State Governments, typically through the state Boards of Pharmacy, have traditionally been the primary regulators of all pharmacy practices.

In the wake of the meningitis outbreak that was caused by injections of a contaminated steroid produced by New England Compounding Center (a compounding pharmacy whose activities appeared to be more akin to a drug manufacturer than a traditional pharmacist), questions have been raised as to the appropriate role for Federal and State regulators for the sector.

Historically, any federal regulatory role for the compounding pharmacy sector has been resisted by the industry. In 2007 when legislation was again circulated to clarify the FDA role in overseeing the safety of compounding pharmacies, it was immediately denounced[15] by trade associations representing the sector.

And a 2011 press release[16] about a court's decision in a veterinary compounding case that was issued by the International Academy of Compounding Pharmacists stated that the court said "what IACP has said for years – the FDA does not have jurisdiction over the traditional practice of pharmacy compounding. That is the sole authority of the state Boards of Pharmacy."

Rep. Markey's staff examined each state Board to determine the degree to which individual States oversee the safety of compounding pharmacy practices, enforce the type of safety-related matters raised by the New England Compounding Center case, and enable the public to search their websites in order to have an understanding of particular pharmacies' compliance records. Table 1 provides a summary of the results of this examination. Rep. Markey's staff examined each state Board of Pharmacy's website, and then additionally followed up with phone calls to each state Board to verify any findings that enforcement actions were not easily made available by a Board to the public.

Rep. Markey's staff analysis found that state Boards of Pharmacy do not, as a general rule, appear to undertake enforcement actions that relate to the safety or scope of compounding pharmacy practices. Their efforts tend to focus more on compliance with traditional pharmacy licensing, controlled substances and other requirements.

Additionally, unlike the FDA enforcement records (which are all publicly available), state Boards vary considerably in terms of what information they make publicly available as well as in the ease with which that information can be found.

In short, it is impossible to conclude from this review of publicly available materials that the safety of pharmacy compounding activities has been or can be sufficiently assured through the reliance on state Boards of Pharmacy alone. Specifically, the review of the state Board websites found that:

- The publicly available information regarding the enforcement activities of state Boards of Pharmacy typically involve traditional types of violations by individual pharmacies or pharmacists (such as billing violations or failure to have a licensed pharmacist onsite) or the distribution of controlled substances via the falsification of prescription records or other means.
- State enforcement records related to the safety of compounding pharmacy practices were not typically found in the enforcement records reviewed by Rep. Markey's staff. The exceptions to this general finding are listed below, and were found in only six States (Arizona, California, Missouri, New York, North Caroline and Rhode Island).
- Most state Board websites do not allow for keyword searches, preventing members of the public from easily locating or downloading enforcement records associated with particular

pharmacies or infractions associated with 'compounding pharmacies' or particular medications.

- Instead of keyword search capability, state enforcement action records related to pharmacies are often limited to alphabetical or temporal lists or summaries of violations that are not themselves searchable. These lists sometimes do not contain sufficient information to understand the nature of a violation in question. A consumer often has to know the date of the enforcement action or the name of the pharmacy in order to obtain useful information, might not be able to search for all safety violations related to 'compounding pharmacies' or the medication they needed, and might not be able to obtain access to the full record of the state Board's enforcement action absent a written request to the Board. Only four state Board websites (Arizona, California, New Jersey and North Carolina) allowed for full keyword searches and enforcement document downloads.
- In many cases, the information available on the state Board websites is limited. For example, while the Massachusetts state Board (and many others) allows for searches of individual pharmacy licensees, the actual enforcement documents that detail the nature of any complaint and the resolution thereof are only available by request to the Board[17]. Additionally, the Massachusetts state Board (and many others) does not post any records associated with complaints that have been voluntarily resolved or dismissed – so the public would not have been able to learn of the earlier safety problems and 2006 consent decree entered into by New England Compounding Center via a search of the Board website.

Appendix B contains a list of the compounding pharmacy safety-related efforts and enforcement actions taken by state Boards of Pharmacy identified via an examination of the publicly available documents on state Board websites, as verified by phone calls as needed.

State guidance or regulatory documents related to compounding pharmacies were largely omitted from this list, as were enforcement actions taken against New England Compounding Center since the fungal meningitis outbreak began.

Table 1. A Survey of State Pharmacy Board Activities Related to the Safety of Compounding Pharmacies

State	URL	Enforcement actions searchable, documents available?	Compounding Pharmacies Cases Found?
Alabama	http://www.albop.com/	No enforcement action information easily available on the website, and multiple efforts to reach a Board employee by phone were unsuccessful.	NO
Alaska	http://www.commerce.state.ak.us/occ/ppha. htm	There is a link to a "Professional License Search" tool, which provides the licensing status but no other information, and no links to documents. Multiple efforts to reach a Board employee by phone were unsuccessful.	NO
Arizona	http://www.azpharmacy.gov/default.asp	YES	YES
Arkansas	http://pharmacyboard.arkansas.gov/Pages/defaul t.aspx	Disciplinary actions taken between 2003--2007 are posted with links to documents, and keyword searches can be performed on the website. Current disciplinary actions are searchable using type of license (pharmacist, wholesale distributor, etc) and at least one search criterion (last name, city, etc), but the only information yielded is whether a disciplinary action exists. To learn more, one has to request documents from the State Board.	NO
California	http://www.pharmacy.ca.gov/	Yes, searchable list also organized by date with links to documents, starting in 2005	YES
Colorado	http://www.dora.state.co.us/pharmacy/	Yes. Summaries of Board enforcement actions from 2008--12 are provided, but are keyword searchable ONLY if one knows the year/month of the enforcement action. Licensees can also be looked up if their name is known on http://www.dora.state.co.us/registrations/ROD.htm.	NO
Connecticut	http://www.ct.gov/dcp/cwp/view.asp?a=1620&Q=273844&PM=1	To obtain information about a specific pharmacy's displinary action history, one has to send a pre-- addressed, postage--paid envelope with a written request to the Board Administrator.	NO
Delaware	http://www.dpr.delaware.gov/	Enforcement acrions are listed at http://www.dpr.delaware.gov/boards/pharmacy/documents/dispaction.pdf with no description of the violation and no way to obtain specific documents. A keyword search using the name of the	NO

Table 1. (Continued)

State	URL	Enforcement actions searchable, documents available?	Compounding Pharmacies Cases Found?
		pharmacist will lead back to this PDF but not provide any additional information. Additional documentation can be obtained through FOIA.	
DC	http://doh.dc.gov/node/185772	No. One can search to find the license for a currently licensed pharmacist if the pharmacist's name is known. To obtain enforcement history for particular pharmacies, one has to request it and pay a fee to receive the documents.	NO
Florida	http://www.doh.state.fl.us/mqa/pharmacy/	Reports prior to 2006 are listed at http://www.doh.state.fl.us/mqa/enforcement/discipline_reports.html but one has to know the year/month the enforcement action took place in order to search on the name of a pharmacist. For more recent enforcement actions, http://ww2.doh.state.fl.us/finalordernet/ allows pharmacist/pharmacy name searches but no other keyword searches. Some documents are posted online, the rest are available on request from the Board.	NO
Georgia	http://sos.georgia.gov/plb/pharmacy/	No. Lists of actions taken by the State Board are provided at http://sos.georgia.gov/plb/pharmacy/news.asp, but these do not provide details of the violation and there does not appear to be a way to obtain documents or do keyword searches because the State Board's search function is inoperable. To obtain more information, requests for records must be faxed to the Board.	NO
Hawaii	http://hawaii.gov/dcca/pvl/boards/pharmacy	No. An individual license status can be found using http://pvl.ehawaii.gov/ pvlsearch/app, and a company's disciplinary action history can be found at http://web.dcca.hawaii.gov/RICO/RICO_RCNT_NAMES/ShowRICO_RCNT_NAMESTable.aspx. However, neither of these searches lead to copies of documents and provide limited information. Monthly press releases of disciplinary actions provide limited descriptions http://hawaii.gov/dcca/oah/news--releases/	NO

State	URL	Enforcement actions searchable, documents available?	Compounding Pharmacies Cases Found?
Idaho	http://bop.idaho.gov/	No. Enforcement actions are listed on the last page of each *Board Newsletter* http://www.nabp.net/publications/idaho--*board--of--pharmacy--newsletters/ and also through the "Verify a License"* tool on the homepage, but neither provides much detail and full documents are available only through a written request to the Board.	NO
Illinois	http://www.idfpr.com/PROFS/Info/pharm.asp	No. http://www.idfpr.com/News/Disciplines/DiscReport s.asp provides a monthly disciplinary action report that keyword searches do work for if one knows what one is looking for, but to obtain documents a FOIA request is required.	NO
Indiana	http://www.in.gov/pla/pharmacy.htm	It is possible to search by pharmacist name at http://www.in.gov/ai/appfiles/pla--litigation/ and obtain full documents. No full keyword search capability.	NO
Iowa	http://www.state.ia.us/ ibpe/	Enforcment records are generally not available online in a searchable format. The Board posts meeting minutes for the past 2 years, and these include disciplinary actions, but one would have to read through them all. Requests for specific records can be made to the Board.	NO
Kansas	http://www.kansas.gov/pharmacy/	Under the "Legal Division" tab on the Board homepage, http://www.kansas.gov/pharmacy/Disciplinary%20C ases.htm lists all enforcement actions against pharmacists or pharmacies by name from 1989 to 2012. Documentation is provided, but one has to know which pharmacist one is looking for.	NO
Kentucky	http://pharmacy.ky.gov/	It is possible to search for specific pharmacists/pharmacies at https://secure.kentucky.gov/pharmacy/licenselookup/to learn their license status and if there has been any disciplinary action. Some documents are available online, others need to be requested from the Board.	NO
Louisiana	http://www.pharmacy.la.gov/	Keyword searches lead to lists of enforcement actions taken in monthly Board meetings, but full documents are not available. Can also look up license status online, but no details or documentation provided absent a request to the Board.https://secure.pharmacy.la.gov/Lookup/LicenseLookup.aspx	NO

Table 1. (Continued)

State	URL	Enforcement actions searchable, documents available?	Compounding Pharmacies Cases Found?
Maine	http://www.maine.gov/pfr/professionallicensing/professions/pharmacy/index.htm	http://pfr.informe.org/almsonline/almsquery/welcome.aspx?AspxAutoDetectCookieSupport=1 provides links to full records, but these are not keyword searchable, so one has to know the name of the pharmacy one is looking for.	NO
Maryland	http://dhmh.maryland.gov/pharmacy/SitePages/ Home.aspx	http://dhmh.maryland.gov/pharmacy/SitePages/for mal--disciplinary---actions.aspx can be searched using the name of the pharmacist but no other keywords.	NO
Massachusetts	http://www.mass.gov/eohhs/provider/licensing/ occupational/pharmacy/	Website allows for searches of individual pharmacy licensees, but the actual enforcement documents that detail the nature of any complaint and the resolution thereof are only available by request to the Board	NO
Michigan	http://www.michigan.gov/lara/0,4601,7--154-- 35299_28150_27529_ 27548------,00.html	http://www.michigan.gov/lara/0,4601,7--154-- 35299_28150_27529---43008----,00.html provides lists of enforcement actions for all professions, but these are not searchable and do not include links to full documentation. To obtain documents about a particular case, a FOIA must be filed.	
Minnesota	http://www.pharmacy.state.mn.us/	Keyword searches lead to summaries of enforcement actions contained in Board meeting newsletters or minutes, but these do not contain full documentation regarding the nature of the complaint.	NO
Mississippi	http://www.mbp.state.ms.us/mbop/pharmacy.ns f	All one can search for is whether a specific pharmacist has a license. There is no posting of disciplinary information at all.	NO
Missouri	http://www.pr.mo.gov/ pharmacists.asp	There is a link to disciplinary actions with full documentation, but no keyword searching is available.	Board conducted annual tests of some compounded medications for potency and sterility and the resulting reports are posted.

State	URL	Enforcement actions searchable, documents available?	Compounding Pharmacies Cases Found?
Montana	http://bsd.dli.mt.gov/license/bsd_boards/pha_board/board_page.asp	To find out if a pharmacy has been subject to disciplinary action, one can search on the name of the pharmacy or license number. No links to full documentation.	NO
Nebraska	http://dhhs.ne.gov/publichealth/Pages/crl_medical_pharm_pharmlic	There are lists containing cursory summaries of the past 3 months' worth of disciplinary actions, but no full documents. One can also look up	NO
	_pharmindex.aspx	individual pharmacy licenses, and these do link to disciplinary documents if any exist.	
Nevada	http://bop.nv.gov/	There is no list of disciplinary actions on the website. In order to acquire information, a request must be submitted to the Board for review (http://bop.nv.gov/services/Disciplinary_Informatio n/). Requestor must have either a case number, license number or business name in order to submit the request and pull records	NO
New Hampshire	http://www.nh.gov/pharmacy/	List of enforcement actions taken since 2009 is at http://www.nh.gov/pharmacy/licensing/documents/ Board_Discipline_01--2009--to--10--2012.pdf, but that contains insufficient information to understand the nature of the infractions. To get further details/documentation, one must submit a request in writing, with a self--addressed stamp and $5 to the Board.	NO
New Jersey	http://www.nj.gov/lps/ca/pharm/	Keyword searches for pharmacy name or words within the complaint are available under the "Board Actions" tab on the website.	NO
New Mexico	http://www.rld.state.nm.us/boards/Pharmacy.as px	A list of enforcement actions with links to fulldocumentation is provided on the website. Can search this by the name of the pharmacist, but keyword searches on the main Board website do not yield results.	NO
New York	http://www.op.nysed.gov/prof/pharm/	Searches for pharmacy name, medication or other keyword lead to summaries of enforcement actions taken available at http://www.op.nysed.gov/opd/rasearch.htm, but no full documents are available absent a request to the Board.	YES
North Carolina	http://www.ncbop.org/	Yes, full enforcement records are available and keyword searches can be performed.	YES

Table 1. (Continued)

State	URL	Enforcement actions searchable, documents available?	Compounding Pharmacies Cases Found?
North Dakota	http://www.nodakpharmacy.com/	If a specific pharmacist's name or license number is known, it is possible to search to see whether any actions have been brought against them at https://www.nodakpharmacy.com/verify.asp, but to obtain documentation a request has to be submitted to the Board.	NO
Ohio	http://www.pharmacy.ohio.gov/	Enforcement actions are summarized in Board meeting minutes at http://www.pharmacy.ohio.gov/Pubs/Minutes.aspx, and while keyword searches do yield these, one has to read the entire document to find brief summaries of information. No enforcement documents are available.	NO
Oklahoma	http://www.ok.gov/ OSBP/	In order to acquire case information, a written request, including the license number and pharmacy name, must be submitted to the Board. Summaries of enforcement information can also be found by searching board meeting minutes (http://www.ok.gov/OSBP/Minutes/index.html)	NO
Oregon	http://www.oregon.gov/Pharmacy /pages/index.a spx	In order to acquire case information, a written request, including the license number and pharmacy name, must be submitted to the Board. General information can also be found by searching through Board meeting minutes but these are not locatable using keyword searches. (http://www.oregon.gov/pharmacy/Pages/Meetings.aspx)	NO
Pennsylvania	http://www.dos.state.pa.us/portal/ server.pt/com munity/state_board_of_pharmacy /12519	There is a list of discipline action summaries organized by date, and a keyword search on the Board's main page will turn up relevant enforcement action summaries. Full documentation is not provided absent a request to the Board. http://www.portal.state. munity/disciplinary_actions/12528	NO
Rhode Island	http://www.health.ri.gov/licenses/ healthcare/in dex.php#pharmacy	A searchable database is at http://www.health. and in some cases full records are available. However, it is only possible to search using the name of the pharmacist, and no other keyword searches turn up results.	YES
South Carolina	http://www.llr.state.sc.us/pol/phar macy/	Enforcement actions are available and searchable by pharmacyor pharmacist name only at at http://www.llr.state.?file=FinalOrders.htm	NO

State	URL	Enforcement actions searchable, documents available?	Compounding Pharmacies Cases Found?
South Dakota	http://doh.sd.gov/boards/ pharmacy/	A list of actions can be found at http://doh.sd.gov/Boards/pharmacy/Discipline.aspx but to obtain documents one has to submit a subpoena, court order or some other authorized request.	NO
Tennessee	http://health. shtml	Enforcement records are organized by month at http://health and are searchable using the name of the pharmacy or pharmacist, and http://health.state.tn.us/Licensure/Default.aspx can also be used to look up individual licenses. Some documents are available when an enforcement action is located, but one has to know the name of the pharmacist.	NO
Texas	http://www.tsbp.state.tx.us/	Keyword searches yield relevant displinary action reports, but full reports only go back to May 2011 at http://www.tsbp.state though past reports are in the process of being added to the website. http://www.tsbp.state.tx.us/dbsearch/default.asp allows individual pharmacists to be looked up to determine whether an enforcement action has been taken, but a written request is required to obtain records.	NO
Utah	http://www.dopl.utah.gov/licensin g/pharmacy.ht ml	There is no list of disciplinary actions taken against pharmacies. http://www.dopl.utah.gov/orders/index.html allows for searches if the license name or number is known.	NO
Vermont	http://vtprofessionals.org/opr1/ph armacists/	Conduct decisions and full documents are available, but not searchable, so one has to know the name of the pharmacy or pharmacist in question.	NO
Virginia	http://www.dhp.virginia.gov/phar macy/	Case decisions are listed, along with some documents, and can be searched using the name of the pharmacy or pharmacist.	NO
Washington	http://www.doh.wa.gov/Licenses PermitsandCerti ficates/ProfessionsNewReneworU pdate/Pharmac y.aspx	There is no list of disciplinary actions, but a search for specific pharmacy or pharmacist names will link to full records if those records are publicly available.	NO

Table 1. (Continued)

State	URL	Enforcement actions searchable, documents available?	Compounding Pharmacies Cases Found?
West Virginia	http://www.wvbop.com/	There is no list of disciplinary actions on the website. For a $10 fee an enforcement action report for specific pharmacies or pharmacists can be obtained.	NO
Wisconsin	http://drl.wi.gov/board_detail.asp?boardid=46&l ocid=0	A list of disciplinary actions is available, and it can be organized by name, order date and county. It is possible to look up the name of a pharmacy or pharmacist and obtain the status of the license including whether there is a history of enforcement actions, but the documents themselves are not available.	NO
Wyoming	http://pharmacyboard.state.wy.us/default.aspx	There is no search feature on the site. There are very general descriptions of past disciplinary cases in Board meeting minutes, but to obtain documents related to a specific pharmacy or pharmacist, a written request to the Board must be made.	NO

Appendix A. A Timeline of Media Reports and FDA Enforcement Actions on Compounding Pharmacies

1. ***June 2001*** – An outbreak of an hospital-acquired bacterial infection occurred at an ambulatory surgery center in California. The outbreak was determined to be caused by a contaminated batch of betamethasone, a steroid solution that was compounded at a community pharmacy and delivered as an injection into the spine of patients. There were a total of 11 cases of infection, five cases of meningitis (of which three died), five cases of epidural abscesses, and one hip infection.[18]
2. ***July 2001*** – The FDA sent a warning letter[19] to Professional Compounding Centers of America, Inc. in Texas because of violations of good manufacturing practices associated with the repackaging of bulk ingredients for use by compounding pharmacies. These violations included a failure to ensure that penicillin-free antibiotics were kept free of contamination with penicillin (and vice-versa), and selling ingredients that had been withdrawn from the market due to safety or efficacy reasons to pharmacies.
3. ***September 2001*** – The FDA sent a warning letter[20] to IV Systems in Illinois because a 2000 inspection of its Texas facility revealed that the facility was repackaging and reselling drugs without the proper labels, its activities revealed deviations from good manufacturing practices, and it was not registered as a pharmacy in Texas.
4. ***October 2001*** – The FDA sent a warning letter[21] to Unique Pharmaceuticals, Ltd. in Texas because the company was operating more like a manufacturer than as a pharmacy, was making copies of commercially available drugs, and its activities had numerous violations of good manufacturing practices.
5. ***January 2002*** – A link was discovered between two deaths of women using prescription- strength numbing cream before laser hair removal. Blanca Bolanos (age 25) died November 1, 2004 after being hooked up to respirator for two years after using the cream, having a seizure, and falling into a coma. Shiri Berg (age 22) died January 25, 2002 under almost identical circumstances – she was headed to Premier Body, a laser hair-removal clinic in North Raleigh. Premier Body closed in February 2005 with plans to liquidate. The women were

given the prescription-strength cream to use outside the clinic. Neither woman had prescriptions for the cream, which was given to them by nonmedical employees at the hair-removal clinics. North Carolina medical and pharmacy boards were investigating Berg's death, as was the the FDA. The creams were both found to have been compounded at a much greater potency than recommended, causing an overdose of the medication.[22]

6. ***April 2002*** – The FDA sent a warning letter[23] to the Compounding Pharmacy in Illinois because it was manufacturing nicotine lollipops and other products that had ingredients that were not FDA-approved, without patient prescriptions, without registering as a manufacturer and without adequate labeling.
7. ***September 2002*** – The FDA sent a warning letter[24] to Med-Mart Pulmonary Services in California because its investigation determined that the company's activities exceeded the scope of the regular course of the practice of pharmacy, and that the company's processes did not conform to good manufacturing processes, including evidence of microbial contamination of its products.
8. ***October 2002*** – An investigation into the meningitis death of a 77-year-old woman who died ten weeks after getting spinal injection at hospital pain clinic implicated Urgent Care Pharmacy, a compounding pharmacy in South Carolina.[25] The pharmacy produced an injectable steroid pain reliever known as methylprednisolone that was contaminated with fungus. Shipments of the infected medicine were traced to North Carolina, South Carolina, Illinois, Indiana, Kentucky, Louisiana, Mississippi, New Hampshire, Massachusetts, Virginia, and Connecticut. North Carolina officials later confirmed three other cases.[26]
9. ***April 2003*** – After receiving several complaints, the FDA investigated and inspected Lee Pharmacy, Inc. of Fort Smith, Arkansas. This pharmacy was responsible for compounding undisclosed amounts of contaminated injectable methylprednisolone acetate. The medication is used to treat joint pain. The pharmacy issued a voluntary recall of the drug in question as well as other medications from the pharmacy amid questions from the FDA regarding the products sterility and potency. Neither federal nor state officials had reported illnesses from the recalled drugs.[27]
10. ***April 2003*** – Kansas City pharmacy (Med 4 Home), owned by Lincare Holdings, Inc. of Clearwater, mishandled the recall of

contaminated drugs, blocked a state inspector from entering its facility, and destroyed records. On March 10, 2003, the Missouri board of health received a temporary restraining order against Med 4 Home, its chief pharmacist, and Lincare's president John P. Byrnes. The injunction barred the pharmacy from compounding and dispensing drugs. Trouble began when a quality assurance test in January 2003 showed bacterial contamination in at least two batches of Albuterol/Ipratropium (compounded inhalant solution for chronic lung diseases). More than 19,000 patients nationwide were estimated to have received the contaminated drug before contamination was discovered. [28]

11. ***May 2003*** – The FDA sent a warning letter[29] to Carneys Drugs in New Hampshire because the company compounded Fentanyl oral lozenges, which the FDA considered to be copies of commercially available drugs.
12. ***October 2003*** – The FDA sent a warning letter[30] to Plum Creek Pharmaceuticals, Inc. in Texas because of the public health risks associated with the compounding of lollipops that contain Fentanyl, Naloxone and Midazolam for use as pain medication to cancer patients. According to the letter, "there have been reports of serious adverse effects, including death, due to accidental pediatric exposure to the commercially available Fentanyl lollipop product in doses comparable to the doses being made available" by the company, but the company did not provide for proper directions of use and childproofing.
13. ***December 2003*** – The FDA sent a warning letter[31] to Custom Compounding Centers in Arkansas because its activities were found to exceed the scope of typical pharmacy activities.
14. ***January 2004*** – The FDA sent a warning letter[32] to Monserrat Pharmaceuticals, Inc. in Puerto Rico because its investigation has determined that its activities exceeded the scope of the regular course of the practice of pharmacy, that it was making copies of commercially available drugs, and because the FDA found numerous violations of good manufacturing practices
15. ***January 2004*** – The FDA sent a warning letter[33] to White Lake Pharmacy in Michigan because it was selling nicotine lollipops without prescriptions in a facility not licensed as a manufacturing facility, and was using advertising that made medical claims about the products' use.

16. ***June 2004*** – The FDA issued a warning letter[34] related to California's Spectrum Chemicals and Laboratory Products' New Jersey manufacturing facility because inspections revealed that the facility was providing to compounding pharmacies the drug domperidone, about which there had been reports of cardiac problems and sudden deaths in patients receiving some intravenous forms that had been withdrawn in several countries. The FDA also sent warning letters to several compounding pharmacies known to have been compounding the substance as well as other substances that FDA warned them to cease compounding (Peoples Pharmacy[35], Inc. in TXexas Drugs Are Us[36] in New Jersey, Axium Healthcare Pharmacy in Florida[37]). At the same time, the FDA warned[38] breastfeeding women not to use domperidone to increase their milk supply.
17. ***July 2004*** – The FDA sent a warning letter[39] to Gentere, Inc. in Ohio because the company, which was not registered with the FDA, manufactured large quantities of injectable drugs without valid prescriptions, and inspections additionally found significant deviations from good manufacturing practices related to its sterile compounding activities.
18. ***September 2004*** – The FDA sent a warning letter[40] to Delta Pharma Inc. in Mississippi because it was found to be more consistent in size and production volume with a drug manufacturer.
19. ***October 2004–*** An outbreak of hepatitis C in Maryland was linked to blood contamination of radiopharmaceutical agent used for myocardial perfusion studies. The original contamination was traced to breaches in aseptic technique at a nuclear pharmacy. In total, 16 patients developed acute hepatitis C infection after undergoing myocardial perfusion studies on the same day at three unaffiliated outpatient clinics. The 16 patients were the only ones injected with the compounded tracer drawn from a single vial prepared at one pharmacy and delivered to the three clinics. [41,42]
20. ***December 2004*** – The FDA sent a warning letter[43] to Lincare, Inc., and Reliant Pharmacy Services, Inc. in Florida because it determined that its activities were akin to that of a drug manufacturer and not a pharmacy.
21. ***December 2004*** – The FDA sent a warning letter[44] to Respi Care Group of Puerto Rico because the FDA's investigation found the company's activities were more akin to that of a drug manufacturer, it produced copies of commercially-available drugs, and it was

manufacturing with "virtually no regard to the current good manufacturing practice", including the identification of major problems related to sterile compounding such as operating in a visibly dirty facility.

22. ***March 2005*** – The FDA issued a nationwide alert concerning contaminated, compounded magnesium sulfate solution used in an IV bag that caused five cases of sepsis in a New Jersey hospital. The medicine was manufactured by PharMEDium Services of Houston, Texas. A South Dakota patient treated with the drug died of sepsis. The product is frequently administered intravenously to patients undergoing cardiac surgery. On April 8, 2005, PharMEDium Services recalled all strengthens of its 50 ml admixtures of Magnesium Sulfate solution. The company also voluntarily ceased production and distribution of the product until it could determine and correct the source of the problem.[45] In April 2007, the FDA sent a warning letter[46] to PharMEDium regarding activities at its TX and MS facilities that was in part related to this alert.
23. ***March 2005*** – The FDA sent a warning letter[47] to Palace Pharmacy in Wyoming because of the company's compounding of drugs containing domperidone (about which there had been reports of cardiac problems and sudden deaths in patients receiving some intravenous forms that had been withdrawn in several countries).
24. ***June 2005*** – The FDA sent a warning letter[48] to Pragmatic Materials, Inc. in Ohio because of violations in good manufacturing practices when it repackaged bulk ingredients for use by compounding pharmacies, including providing non-prescription grade ingredients that the company labeled as prescription grade,
25. ***July 2005*** – The FDA sent a letter[49] to Cape Drugs in Maryland because of the company's compounding of drugs containing domperidone (about which there had been reports of cardiac problems and sudden deaths in patients receiving some intravenous forms that had been withdrawn in several countries).
26. ***August 2005*** – Custom Rx Compounding Pharmacy in Richfield, Minnesota initiated a nationwide recall of Trypan Blue Ophthalmic Solution because it was contaminated with bacteria that could lead to serious injury – potentially blindness – if applied to the eyes. The drug was distributed to hospitals and clinics in MD, MN, IL, NE, ND, MI, DC, and PA and was intended for use during cataract surgery. The pharmacy voluntarily recalled the product based on two reports of lost

vision associated with use of the product as reported by Centers for Disease Control.[50]

27. ***September 2005*** – The FDA notified healthcare professionals and hospitals of a product recall involving all injectable compounded products made by Central Admixture Pharmacy Service (CAPS) of Lanham, Maryland. The products were recalled because of sterility concerns that were discovered after several patients that received the CAPS medication developed a severe systemic inflammatory response. Four deaths and eleven other severe reactions were suspected to have resulted from these problems[51]. CAPS distributed the affected injectable products to hospitals in MD, DE, DC, and VA.[52] The FDA sent a warning letter[53] to the company in February 2006 telling it to remedy deficiencies related to its sterile compounding activities in its facilities in AL, MD, PA, and MO as well as other deficiencies in its CA facility.
28. ***November 2005*** – The FDA sent a warning letter[54] to Spectrum Chemicals and Laboratory Products, Inc in Arizona because its New Jersey facility was repackaging bulk domperidone (about which there had been reports of cardiac problems and sudden deaths in patients receiving some intravenous forms that had been withdrawn in several countries) and polidocanol (with reported adverse events including infections at the treated site and reversible cardiac arrest after polidocanol sclerotherapy) and reselling it to pharmacies for compounding. This was the second such letter to this company (see June 2004).
29. ***December 2005*** – The FDA sent a warning letter[55] to Samson Medical Technologies, Inc. in New Jersey because its inspection found that the company was marketing antibiotics labeled as being usable in compounding by pharmacies even though that was not their intended use. The products were supposed to be reconstituted and then dispensed to patients in a hospital setting.
30. ***February 2006*** – Following a 2005 inspection, the FDA sent a warning letter[56] to Southern Meds Joint Venture, LLC in Mississippi, telling it to cease its actions which included compounding of drugs in volumes that was not "consistent with traditional pharmacy compounding operation" and actions which included significant violations of good manufacturing practices related to the prevention of contamination of sterile products.

31. ***August 2006*** – The FDA notified[57] consumers and healthcare professionals that RoTech Healthcare[58], Inc. in Florida, CCS Medical[59] in Florida, and Reliant Pharmacy Services in Florida were manufacturing and distributing unapproved compounded inhalation drugs nationwide for use in patients with asthma, emphysema, bronchitis, and cystic fibrosis. The FDA warned the company that they were mass-producing unapproved drugs and were operating outside of the scope of a typical pharmacy.
32. ***September 2006*** – The FDA sent a warning letter[60] to Hawkins, Inc. in Minnesota because it was found to be repackaging bulk ingredients and re-selling them to compounding pharmacies, and the FDA found significant deviations from good manufacturing practices.
33. ***October 2006*** – The FDA sent a warning letter[61] to Wedgwood Village Pharmacy in New Jersey because the pharmacy was found to have been compounding copies of commercially-available drugs and compounding drugs in volumes that were in excess of traditional compounding pharmacy practices.
34. ***October 2006*** – The FDA sent a warning letter[62] to Pharmacy Creations in New Jersey because it was compounding domperidone (about which there had been reports of cardiac problems and sudden deaths in patients receiving some intravenous forms that had been withdrawn in several countries) and polidocanol, neither of which were FDA-approved active ingredients, as well as adenosine-5-monophosphate, which had been withdrawn from the market because it was neither safe nor effective, and was also compounding copies of commercially available products.
35. ***November 2006*** – The FDA sent a warning letter[63] to Health Dimensions Inc. in Michigan because it was compounding domperidone, (which was not FDA-approved and about which there had been reports of cardiac problems and sudden deaths in patients receiving some intravenous forms that had been withdrawn in several countries). A second (name redacted but possibly polidocanol) drug, also not FDA-approved and known to have adverse events including deep venous thromboses, necrosis, and ulceration at the treated site and reversible cardiac arrest, was found to be compounded by the pharmacy.
36. ***December 2006*** – The FDA sent a warning letter[64] to Spoonamore Drug Co in Kentucky because it was producing domperidone capsules (which was not FDA-approved, and about which there had been

reports of cardiac problems and sudden deaths in patients receiving some intravenous forms that had been withdrawn in several countries), progesterone capsules (which FDA believed was a copy of a commercially available product), testosterone five percent gel which FDA believed was a copy of a commercially-available product), and nicotine lollipops (which the FDA said lacked the required warning labels and may have been compounded using ingredients that were not FDA-approved).

37. ***December 2006*** – *The* FDA sent warnings[65] to five pharmacies that were compounding topical anesthetic creams that were marketed to the general public rather than in response to specific patient prescriptions. The FDA noted that these creams "can cause grave reactions including seizures and irregular heartbeats. Two deaths have been connected to compounded topical anesthetic creams made by Triangle Compounding Pharmacy[66] (North Carolina) and University Pharmacy[67] (Utah), two of the five pharmacies receiving warning letters." Other non-fatal reactions to the creams had also been documented. The three other pharmacies included New England Compounding Center[68] (Massachusetts), Custom Scripts Pharmacy[69] (Florida), and Hal's Compounding Pharmacy[70] (California). The FDA found additional problems with other drugs that were being compounded by these pharmacies.
38. ***January 2007*** – The FDA sent a warning letter[71] to Kalchem, International, an Oklahoma company that was found to be distributing domperidone (which was not FDA- approved, and about which there had been reports of cardiac problems and sudden deaths in patients receiving some intravenous forms that had been withdrawn in several countries) to pharmacies for compounding.
39. ***March 2007*** – The FDA sent a warning letter[72] to ComputeRx/ Broncho-Dose, Ltd in Connecticut because it was found to have been compounding inhalation drugs in quantities that exceed traditional pharmacy practices and across 12 states. Additionally, the pharmacy ended up recalling one of its formulations because it was found to be less potent than the label advertised.
40. ***April 2007*** – The FDA sent a warning letter[73] to PharMEDium Services regarding its facilities in Texas and Mississippi. The letter related to 2005 problems with the sterility of its compounded magnesium sulfate solution, some of which caused five cases of sepsis in a New Jersey hospital and a death of a patient in South Dakota. On

April 8, 2005, PharMEDium Services recalled all strengthens of its 50 ml admixtures of Magnesium Sulfate solution. The letter also related to an adverse event reportedly associated with an epidural injection made in the company's Mississippi facility that was supposed to contain Fentanyl/Bupivacaine but actually contained Morphine Sulfate. The patient in question showed signs and symptoms of decreased consciousness, hypoxia, and hypotension. The FDA found additional labeling and manufacturing problems associated with the facilities.

41. ***April 2007*** – A measuring error at ApotheCure, a Texas compounding pharmacy, shipped 31 vials of mislabeled Colchicine into Oregon that was eight times more concentrated than the recognized standard level. Colchicine, which has not been approved by the FDA, has been used since the 1950s to treat gout. More recently, it has been used by alternative medicine clinics to treat chronic back pain. In April 2007, at the request of the Texas State Board of Pharmacy, the company issued an immediate recall for all strengths, sizes and lots of this compounded drug that had been sold within the previous year.[74]
42. ***September 2007*** – The FDA sent a warning letter[75] to Med-South Pharmacy Inc. in Alabama following reports of injuries relating to betamethasone acetate/betamethasone sodium phosphate multi-dose injectable drug product. "The complaints included redness, large swollen areas, bruising at the injection site, rash, fever, and cellulitis, with some patients requiring intravenous antibiotics." The FDA also noted that the company's activities exceeded the scope of activities typically conducted by a pharmacy, identified problems with the company's sterile compounding practices, and found the company was making copies of commercially available drugs.
43. ***January 2008*** – The FDA warned[76] that pharmacies that were making claims about the safety and effectiveness of their bio-identical hormone replacement therapy, or BHRT products were unsupported by medical evidence and were considered false and misleading by the agency. The drugs contain hormones such as estrogen, progesterone and estriol, which is an FDA unapproved drug.[77] The pharmacies that received the letters included Panorama Compounding Company[78] in California, Saint John's Medical Plaza Pharmacy in California, Murray Avenue Apothecary[79] in Pennsylvania, Village Compounding Pharmacy[80] in Texas, Pharmacy Compounding Specialties[81] in Texas, Reed's Compounding Pharmacy[82] in Arizona, and Pacifica

Pharmacy[83] in California. The FDA also sent warning letters to American Hormones Inc.[84] in New York and Bellevue Pharmacy Solutions[85] in Missouri that included some of the same concerns, and the FDA also noted that these companies were engaging in activities that exceeded the scope of typical pharmacies' manufacturing practices, and were making copies of commercially-available drugs, and found that other drugs being compounded by the pharmacies were not FDA approved or were known to be associated with adverse reactions.

44. ***March 2008*** – The FDA sent a warning letter[86] to Farmacia La Salud in Puerto Rico, because it was manufacturing, in scales inconsistent with traditional pharmacies, copies of commercially-available inhalation drugs, and that the company's sterility practices were problematic.
45. ***June 2008*** – The FDA sent a warning letter[87] to Newman Inc. in Alabama because the firm was found to have been compounding drugs in quantities that exceed traditional pharmacy practice, and that the health claims the company made about the drugs were false and misleading.
46. ***October 2008*** – The FDA sent a warning letter[88] to Aerosol Science Laboratories Inc. in California because of false and misleading claims made about the company's nasal aerosol drugs.
47. ***November 2008*** – The FDA sent a warning letter[89] to Steven's Pharmacy in California because the firm was found to have been compounding drugs in quantities that exceeded traditional pharmacy practice, as well as drugs that were copies of commercially- available products. The FDA additionally noted that the topical anesthetics that the company was compounding contained high doses of drugs that were not approved for topical use and which had been associated with serious cases of systemic toxicity, and that the labels associated with the products did not contain necessary warnings. The company was also found to be compounding drugs that were not FDA-approved.
48. ***December 2008*** – The FDA sent a warning letter[90] to Civic Center Pharmacy in Arizona because the company was found to compound a hormone therapy drug containing estriol, which was not FDA-approved, and that it also compounded copies of commercially available drugs.

49. ***December 2008*** – FDA sent a warning letter[91] to Kreitchman PET Center in New York because of deviations from good manufacturing practices associated with the compounding of PET drugs.
50. ***January 2009*** – The FDA sent a warning letter[92] to Cyclotron Center of NE Florida because its activities were found to deviate from good manufacturing practices for PET drugs, including repeat violations related to the sterility of their processes that were also found in 2005.
51. ***September 2009*** – The FDA sent a warning letter[93] to Hopewell Pharmacy and Compounding Center in New Jersey because it was found to be compounding Sodium Tetradecyl Sulfate (STS) Injection, which was found to have been contaminated.
52. ***April 2010*** – The FDA sent a warning letter[94] to J & F International Inc., dba Alexandria Medical Arts Pharmacy & Compounding Laboratory in Virginia because it was found to be making domperidone (which was not FDA-approved, and about which there had been reports of cardiac problems and sudden deaths in patients receiving some intravenous forms that had been withdrawn in several countries).
53. ***April 2010*** – The FDA issued a warning[95] to six U.S. based medical spas and a company in Brazil for making false or misleading statements about drugs they claimed would eliminate fat in a procedure called "lipodissolve," or for otherwise misbranding lipodissolve products. The letters were sent to Pure Med Spa[96] in Florida, All About You Medspa[97] in Indiana, Ininhealth[98] in Minnesota, Spa 35[99]in Idaho, Monarch Med Spa[100] in Pennsylvania, Medical Cosmetic[101] in Maryland, and Ziomed and Mesoone in Brazil.
54. ***July 2010*** – The FDA sent a warning letter[102] to Med Prep Consulting, Inc. in New Jersey, a company that repackaged sterile drug products for other entities without a patient prescription and in a manner that was more consistent with a manufacturer than a pharmacist.
55. ***November 2010*** – The FDA sent a warning letter[103] to VanderVeer Center in Oregon because it was making false or misleading claims about the safety and effectiveness of its Lipodissolve products, which were claimed to eliminate fat.
56. ***March 2011*** – The FDA sent a warning letter[104] to Proportional Technologies, Inc. in Texas because of significant violations related to the manufacturing of PET drugs, including practices related to assuring the sterility of the drugs.

57. ***April 2011*** – The Alabama Department of Public Health confirmed that bacteria from a Birmingham, Alabama pharmacy sickened 19 people and killed nine others, when patients in a hospital were given an intravenous nutritional supplement. Samples gathered from the Meds IV's Oxmoor Road compounding facility tested positive for lethal bacteria introduced to the pharmacy's nutritional supplement when tap water used to clean a container was used for mixing the drug.[105] A number of other cases of illness in the preceding months were believed to have been linked to the compounding facility.
58. ***August 2011*** – *The New York Times* reported that four patients in a Tennessee Veterans hospital suffered complications, including brain damage, and blindness from eye injections of Avastin, which had been contaminated by bacteria when the drug was compounded by the pharmacy in the VA hospital in Nashville. Avastin was used to treat macular degeneration, the most common cause of blindness in older Americans. The FDA has not approved Avastin for eye injections. Lucentis, the drug approved by the FDA for macular degeneration, costs about $2,000 for a single-dose vial. Avastin only costs around $50 a dose, and according to a study published by the New England Journal of Medicine, it is equally effective in fighting the disease. Hospitals turn to Avastin as a cheaper alternative to Lucentis. However, Avastin does not come in single-dose vials for eye injections, so hospital pharmacies have to compound Avastin for eye injections from packaging intended for intravenous use. A similar situation in the Miami, Florida area where a compounding pharmacy repacked Avastin and shipped to eye clinics in the area left 12 people with lost of severely impaired vision. [106] The FDA issued a warning to health care professionals warning of the problem[107].
59. ***July 2012*** – The FDA sent a warning letter[108] to Franck's Lab, Inc. dba Franck's Compounding Lab in Florida because some of its injectable products were found to be contaminated, following reports of patients who developed eye infections. Inspections revealed unsanitary conditions and multiple bacterial and fungal species at various locations in the facility.
60. ***July 2012*** – The FDA sent a warning letter[109] to Infupharma in Florida because it identified significant violations of Current Good Manufacturing Practice including those related to sterile compounding, its activities went beyond the scope of traditional

pharmacy activities, and because inspections found bacterial contamination in samples of Avastin the company was repackaging.

APPENDIX B. A TIMELINE OF STATE PHARMACY BOARD ACTIONS ON COMPOUNDING PHARMACIES

Arizona

October 2003 – The State of Arizona filed a complaint[110] and later entered into a consent agreement[111] with the Apothecary Shop of Phoenix after two patients received compounded prescriptions for liothyronine that caused their blood levels of thyroid hormone to rise to 800 times the normal range. The pharmacy turned out to have provided the patients with raw materials rather than a diluted compounded drug.

January 2004– The State of Arizona entered into a consent agreement[112] with pharmacist James R. Bataoel after he compounded medications with the incorrect dose.

May 2008 – The State of Arizona entered into a consent agreement[113] with pharmacist Joseph Chen because the pharmacy had many long-expired drugs on-hand and the pharmacist said that it was too busy to pay attention to expiration dates when compounding.

July 2010 – The State of Arizona entered into a consent agreement[114] with pharmacist Daniel Hoffman because of more than one failure to compound a prescribed medication correctly, which resulted in patients experiencing side effects.

July 2010 – The State of Arizona entered into a consent agreement[115] with pharmacist Vicki Graves because of an error made compounding thyroid medication that resulted in four patients being given doses 10 times more potent than what was prescribed, resulting in at least one pregnant patient experiencing side effects.

California

April 2008- The State of California denied[116] an application by an individual to be a Pharmacy Technician because he had previously been licensed as a pharmacist but had been disciplined because he and his colleagues had (among other things) compounded bacterially-contaminated betamethasone, a steroid, that was sent to six different health care facilities. According to the denial, 38 patients received the medication in May 2001, 13 were hospitalized and three died.

December 2009- As part of a disciplinary action[117] related to the distribution of controlled substances, Advanced Physician Solutions dba Advanced Compounding Pharmacy of CA was also found to have violated record-keeping and procedural requirements related to sterile compounding and had mislabeled medications they sold.

February 2011 – The State of California brought a complaint[118] against Superior Medical Supply in CO because it was (among other things) selling misbranded drugs and was additionally distributing drugs produced by Advanced Compounding Pharmacy (see December 2009).

March 2012- *The* State of California brought a complaint against The Compounding Shop[119] in CA. The complaint alleged (among other things) that the company dispensed medications with labeled potencies that were different from what was labeled, and used non-sterile ingredients to compound what was advertised to be sterile products.

September 2012- The State of California brought an enforcement case against Franck's Compounding Pharmacy in FL following reports in March 2012 of outbreaks of fungal infections that resulted from contamination of medications compounded by Franck's. According to the report, "the total number of doses, prescriptions, and/or patients affected is not known, but at least twenty (20) confirmed and probable cases (7 confirmed, 13 probable) of fungal infection resulting from BBG compounded by Respondent, and at least thirteen (13) such cases (II confirmed, 2 probable) of infections resulting from TMC compounded by Respondent, were identified in seven (7) states. Up to seventeen (17) of these cases were in California."

Missouri

2003 – In 2003, the State began to conduct annual tests[120] of some compounded drugs for potency, sterility and some evidence of bacterial contamination. Unsatisfactory results were reported to the compounding pharmacies responsible who were also asked to prepare corrective action plans. The results showed that between 11.6-25.2% of drugs that were tested had unsatisfactory results. The reports list the drugs that were found to be unsatisfactory results, but didn't indicate which pharmacies were responsible.

New York

February 2001 – The State of New York entered into consent orders[121] with several pharmacists that allowed the compounding and mixing of drugs by unlicensed pharmacy technicians.

December 2004 – The State of New York entered into a consent order[122] with pharmacist Kevin Ferry for numerous violations, including delegating to a pharmacy technician the responsibility of compounding a patient-specific intravenous solution.

North Carolina

January 2006 – The State entered into a consent order[123] with Professional Home Pharmacy, which had reportedly allowed a non-pharmacist to compound respiratory medications without proper supervision by a pharmacist.

March 2007 – The State entered into a consent decree[124] with The Pharmaceutical Center for numerous infractions, including a case where a drug compound had been improperly made and a failure to maintain adequate records for its compounded drugs.

April 2010 – The State entered into consent decrees[125, 126] with pharmacists William B. Cheek and Michael Rogers who were found to have compounded medications for patients without prescriptions on more than 100 occasions.

April 2011 – the State entered into a consent decree[127] with pharmacist Jewel Freeman who dispensed a 9 percent sodium chloride solution in place of a 0.9% solution that was prescribed to a pediatric oncology patient who then died following administration of the drug.

May 2011 – The State entered into consent decrees[128,129] with pharmacist Bradley Weaver and Deep River Drug, which had dispensed a 30-day supply of compounded levothyroxine to a patient in doses that were many times higher than what was prescribed. The patient died, but Mr. Weaver did not report the compounding error to the Board.

February 2012- The State entered into a consent decree[130] with Royal Palm Compounding Pharmacy, which sold compounded drugs directly to physicians who then re-sold them to patients and sold compounded drugs to patients who had prescriptions from physicians in other States in a manner that suggested that the company should have known that the patients were not being examined by the prescribing physicians.

October 2012 – The State announced the formation of a working group on compounding pharmacies and suspended[131] New England Compounding Center's license.

Rhode Island

March 2012 – The State of Rhode Island issued an Order[132] to Apothecare Compounding Solutions because it was in possession of numerous misbranded pharmaceuticals and the compounding area was cluttered and messy.

End Notes

[1] http://www.crs.gov/Products/R/PDF/R40503.pdf
[2] http://iacprx.affiniscape.com/associations/13421/files/U.S._District_Court_Rules_Favorably_for_Pharmacy_Compoun.pdf
[3] Private communication between MA State officials and Rep. Markey's staff
[4] http://www.crs.gov/Products/R/PDF/R40503.pdf
[5] http://www.crs.gov/Products/R/PDF/R40503.pdf
[6] http://www.crs.gov/Products/R/PDF/R40503.pdf
[7] http://www.fda.gov/downloads/AboutFDA/CentersOffices/CDER/UCM118050.pdf

[8] http://www.fda.gov/Drugs/GuidanceComplianceRegulatoryInformation/Pharmacy Compounding/ucm155725.htm
[9] http://www.fda.gov/Drugs/GuidanceComplianceRegulatoryInformation/Pharmacy Compounding/ucm204237.htm
[10] http://www.crs.gov/Products/R/PDF/R40503.pdf
[11] http://www.foxnews.com/health/2012/10/16/how-compounding-pharmacies-rallied-patients-to-fight-regulation/
[12] http://online.wsj.com/article/SB10000872396390444657804578052972230404046.html? Mod =googlenews_wsj
[13] http://www.fda.gov/downloads/iceci/enforcementactions/enforcementstory/enforcementstory archive/ucm091066.pdf
[14] http://drugtopics.modernmedicine.com/drugtopics/Community+Pharmacy/New-bill-on-pharmacy-compounding- stirs-concern/ArticleStandard/Article/detail/414436
[15] http://drugtopics.modernmedicine.com/drugtopics/Community+Pharmacy/New-bill-on-pharmacy-compounding- stirs-concern/ArticleStandard/Article/detail/414436
[16] http://iacprx.affiniscape.com/associations/13421/files/U.S._District_Court_Rules_ Favorably_for_Pharmacy_Compoun.pdf
[17] Private communication between MA State officials and Rep. Markey's staff
[18] http://cid.oxfordjournals.org/content/43/7/831.full.pdf+html
[19] http://www.fda.gov/ICECI/EnforcementActions/WarningLetters/2001/ucm178377.htm
[20] http://www.fda.gov/ICECI/EnforcementActions/WarningLetters/2001/ucm178394.htm
[21] http://www.fda.gov/ICECI/EnforcementActions/WarningLetters/2001/ucm178396.htm
[22] Orlando Sentinel (Florida), February 12, 2005 Saturday, A SECTION; FLORIDA; Pg. A27, 654 words, Yonat Shimron and Michael Easterbrook, the (Raleigh, N.C.) News & Observer
[23] http://www.fda.gov/ICECI/EnforcementActions/WarningLetters/2002/ucm144843.htm
[24] http://www.fda.gov/ICECI/EnforcementActions/WarningLetters/2002/ucm145144.htm
[25] http://vitals.nbcnews.com/_news/2012/10/18/14517177-fungal-meningitis-outbreak-tied-to-steroid-shots-isnt-the- first-reports-show?lite
[26] Kansas City Star (Kansas & Missouri), October 2, 2002, Wednesday, 498 words, MARK MORRIS; DONNA McGUIRE; The Kansas City Star
[27] Kansas City Star, April 11, 2003, Friday, 894 words, MARK MORRIS; DONNA McGUIRE; The Kansas City Star
[28] St. Petersburg Times (Florida), April 18, 2003 Friday 0 South Pinellas Edition, BUSINESS; Pg. 1E, 806 words, KRIS HUNDLEY
[29] http://www.fda.gov/ICECI/EnforcementActions/WarningLetters/2003/ucm147501.htm
[30] http://www.fda.gov/ICECI/EnforcementActions/WarningLetters/2003/ucm147759.htm
[31] http://www.fda.gov/ICECI/EnforcementActions/WarningLetters/2003/ucm147890.htm
[32] http://www.fda.gov/ICECI/EnforcementActions/WarningLetters/2004/ucm146158.htm
[33] http://www.fda.gov/ICECI/EnforcementActions/WarningLetters/2004/ucm146154.htm
[34] http://www.fda.gov/downloads/Drugs/GuidanceComplianceRegulatoryInformation/ EnforcementActivitiesbyFDA/WarningLettersandNoticeofViolationLetterstoPharmaceutica lCompanies/UCM054622.pdf
[35] http://www.fda.gov/downloads/Drugs/GuidanceComplianceRegulatoryInformation/ EnforcementActivitiesbyFDA/WarningLettersandNoticeofViolationLetterstoPharmaceutica lCompanies/UCM054621.pdf
[36] http://www.fda.gov/downloads/Drugs/GuidanceComplianceRegulatoryInformation/ EnforcementActivitiesbyFDA/WarningLettersandNoticeofViolationLetterstoPharmaceutica lCompanies/UCM054620.pdf
[37] http://www.fda.gov/downloads/Drugs/GuidanceComplianceRegulatoryInformation/ EnforcementActivitiesbyFDA/WarningLettersandNoticeofViolationLetterstoPharmaceutica lCompanies/UCM054619.pdf
[38] Endnote note available.

[39] http://www.fda.gov/ICECI/EnforcementActions/WarningLetters/2004/ucm146503.htm
[40] http://www.fda.gov/ICECI/EnforcementActions/WarningLetters/2004/ucm146601.htm
[41] Reuters Health Medical News, October 24, 2006 Tuesday 9:00 PM EST, PUBLIC HEALTH, 499 words
[42] http://jama.jamanetwork.com/article.aspx?articleid=203792
[43] http://www.fda.gov/ICECI/EnforcementActions/WarningLetters/2004/ucm146702.htm
[44] http://www.fda.gov/ICECI/EnforcementActions/WarningLetters/2004/ucm146721.htm
[45] http://www.fda.gov/Safety/MedWatch/SafetyInformation/SafetyAlertsforHumanMedical Products/ucm152363.htm
[46] http://www.fda.gov/ICECI/EnforcementActions/WarningLetters/2007/ucm076357.htm
[47] http://www.fda.gov/ICECI/EnforcementActions/WarningLetters/2005/ucm075333.htm
[48] http://www.fda.gov/ICECI/EnforcementActions/WarningLetters/2005/ucm075449.htm
[49] http://www.fda.gov/ICECI/EnforcementActions/WarningLetters/2005/ucm075482.htm
[50] http://www.fda.gov/ForConsumers/ConsumerUpdates/ucm107836.htm
[51] http://usatoday30.usatoday.com/news/health/2006-08-07-unsterile-drugs_x.htm
[52] http://www.fda.gov/Safety/MedWatch/SafetyInformation/SafetyAlertsforHuman MedicalProducts/ucm153095.htm
[53] http://www.fda.gov/ICECI/EnforcementActions/WarningLetters/2006/ucm075828.htm
[54] http://www.fda.gov/ICECI/EnforcementActions/WarningLetters/2005/ucm075714.htm
[55] http://www.fda.gov/ICECI/EnforcementActions/WarningLetters/2005/ucm075721.htm
[56] http://www.fda.gov/ICECI/EnforcementActions/WarningLetters/2006/ucm075794.htm
[57] http://www.fda.gov/NewsEvents/Newsroom/PressAnnouncements/2006/ucm108709.htm
[58] http://www.fda.gov/ICECI/EnforcementActions/WarningLetters/2006/ucm076025.htm
[59] http://www.fda.gov/ICECI/EnforcementActions/WarningLetters/2006/ucm076026.htm
[60] http://www.fda.gov/ICECI/EnforcementActions/WarningLetters/2006/ucm076059.htm
[61] http://www.fda.gov/ICECI/EnforcementActions/WarningLetters/2006/ucm076147.htm
[62] http://www.fda.gov/ICECI/EnforcementActions/WarningLetters/2006/ucm076148.htm
[63] http://www.fda.gov/ICECI/EnforcementActions/WarningLetters/2006/ucm076182.htm
[64] http://www.fda.gov/ICECI/EnforcementActions/WarningLetters/2006/ucm076191.htm
[65] http://www.fda.gov/NewsEvents/Newsroom/PressAnnouncements/2006/ucm108793.htm
[66] http://www.fda.gov/ICECI/EnforcementActions/WarningLetters/2006/ucm076193.htm
[67] http://www.fda.gov/ICECI/EnforcementActions/WarningLetters/2006/ucm076192.htm
[68] http://www.fda.gov/ICECI/EnforcementActions/WarningLetters/2006/ucm076196.htm
[69] http://www.fda.gov/ICECI/EnforcementActions/WarningLetters/2006/ucm076194.htm
[70] http://www.fda.gov/ICECI/EnforcementActions/WarningLetters/2006/ucm076195.htm
[71] http://www.fda.gov/ICECI/EnforcementActions/WarningLetters/2007/ucm076233.htm
[72] http://www.fda.gov/ICECI/EnforcementActions/WarningLetters/2007/ucm076329.htm
[73] http://www.fda.gov/ICECI/EnforcementActions/WarningLetters/2007/ucm076357.htm
[74] http://www.fda.gov/Safety/MedWatch/SafetyInformation/SafetyAlertsforHumanMedical Products/ucm152108.htm
[75] http://www.fda.gov/ICECI/EnforcementActions/WarningLetters/2007/ucm076516.htm
[76] http://www.fda.gov/NewsEvents/Newsroom/PressAnnouncements/2008/ucm116832.htm
[77] U.S. Food & Drug Administration, January 1, 2008, http://www.fda.gov/Safety/MedWatch/ SafetyInformation/SafetyAlertsforHumanMedicalProducts/ucm110239.htm
[78] http://www.fda.gov/ICECI/EnforcementActions/WarningLetters/2008/ucm1048441.htm
[79] http://www.fda.gov/ICECI/EnforcementActions/WarningLetters/2008/ucm1048443.htm
[80] http://www.fda.gov/ICECI/EnforcementActions/WarningLetters/2008/ucm1048444.htm
[81] http://www.fda.gov/ICECI/EnforcementActions/WarningLetters/2008/ucm1048445.htm
[82] http://www.fda.gov/ICECI/EnforcementActions/WarningLetters/2008/ucm1048446.htm
[83] http://www.fda.gov/ICECI/EnforcementActions/WarningLetters/2008/ucm1048447.htm
[84] http://www.fda.gov/Drugs/GuidanceComplianceRegulatoryInformation/Pharmacy Compounding/ucm155170.htm

[85] http://www.fda.gov/Drugs/GuidanceComplianceRegulatoryInformation/Pharmacy Compounding/ucm155169.htm
[86] http://www.fda.gov/Drugs/GuidanceComplianceRegulatoryInformation/ PharmacyCompounding/ucm155168.htm
[87] http://www.fda.gov/ICECI/EnforcementActions/WarningLetters/2008/ucm1048230.htm
[88] http://www.fda.gov/ICECI/EnforcementActions/WarningLetters/2008/ucm1048081.htm
[89] http://www.fda.gov/ICECI/EnforcementActions/WarningLetters/2008/ucm1048074.htm
[90] http://www.fda.gov/ICECI/EnforcementActions/WarningLetters/2008/ucm1048048.htm
[91] http://www.fda.gov/ICECI/EnforcementActions/WarningLetters/2008/ucm162944.htm
[92] http://www.fda.gov/ICECI/EnforcementActions/WarningLetters/2009/ucm152644.htm
[93] http://www.fda.gov/ICECI/EnforcementActions/WarningLetters/2009/ucm188449.htm
[94] http://www.fda.gov/ICECI/EnforcementActions/WarningLetters/2010/ucm208772.htm
[95] http://www.fda.gov/NewsEvents/Newsroom/PressAnnouncements/2010/ucm207453.htm
[96] http://www.fda.gov/ICECI/EnforcementActions/WarningLetters/2010/ucm207642.htm
[97] http://www.fda.gov/ICECI/EnforcementActions/WarningLetters/2010/ucm213160.htm
[98] http://www.fda.gov/ICECI/EnforcementActions/WarningLetters/2010/ucm207658.htm
[99] http://www.fda.gov/ICECI/EnforcementActions/WarningLetters/2010/ucm207646.htm
[100] http://www.fda.gov/ICECI/EnforcementActions/WarningLetters/2010/ucm207640.htm
[101] http://www.fda.gov/ICECI/EnforcementActions/WarningLetters/2010/ucm207651.htm
[102] http://www.fda.gov/ICECI/EnforcementActions/WarningLetters/2010/ucm222283.htm
[103] http://www.fda.gov/ICECI/EnforcementActions/WarningLetters/2010/ucm235714.htm
[104] http://www.fda.gov/ICECI/EnforcementActions/WarningLetters/2011/ucm286635.htm
[105] http://blog.al.com/spotnews/2011/03/nine_dead_in_alabama_hospitals.html
[106] http://www.nytimes.com/2011/08/31/health/31drug.html
[107] http://www.fda.gov/Drugs/DrugSafety/ucm270296.htm
[108] http://www.fda.gov/ICECI/EnforcementActions/WarningLetters/2012/ucm312645.htm
[109] http://www.fda.gov/ICECI/EnforcementActions/WarningLetters/2012/ucm317190.htm
[110] http://www.azpharmacy.gov/disciplines/pdfs/03-0015%20B.pdf
[111] http://www.azpharmacy.gov/disciplines/pdfs/03-0015.pdf
[112] http://www.azpharmacy.gov/disciplines/pdfs/03-0018.pdf
[113] http://www.azpharmacy.gov/disciplines/pdfs/08-0039.pdf
[114] http://www.azpharmacy.gov/disciplines/pdfs/10-0060.pdf
[115] http://www.azpharmacy.gov/disciplines/pdfs/10-0061.pdf
[116] http://www.pharmacy.ca.gov/enforcement/fy0708/si073163.pdf
[117] http://www.pharmacy.ca.gov/enforcement/accusations/ac083251.pdf, http://www.pharmacy. ca. gov/enforcement/fy0809/ac083251_rph22604.pdf
[118] http://www.pharmacy.ca.gov/enforcement/fy0809/ac083331_osd.pdf
[119] http://www.pharmacy.ca.gov/enforcement/accusations/ac103990.pdf
[120] http://www.pr.mo.gov/pharmacists-compounding.asp
[121] http://www.op.nysed.gov/opd/feb01.htm
[122] http://www.op.nysed.gov/opd/dec04.htm
[123] http://www.ncbop.org/Disciplinary%20Actions%20-%20PHARMACIES/Professional %20Care %20Phcy%207176%201%2006%20CO.pdf#search="compounding pharmacy"
[124] http://www.ncbop.org/Disciplinary%20Actions%20-%20PHARMACISTS/A%20Artemes% 204190%20%20Pharmceutical%20Center%203013.pdf#search="compounding pharmacy"
[125] http://www.ncbop.org/Disciplinary%20Actions%20-%20PHARMACISTS/CheekWilliam 07868. pdf#search="compounding pharmacy"
[126] http://www.ncbop.org/Disciplinary%20Actions%20-%20PHARMACISTS/RogersMichael 06864.pdf#search="compounding pharmacy"
[127] http://www.ncbop.org/Disciplinary%20Actions%20-%20PHARMACISTS/FreemanJewel 10249.pdf#search="compounding pharmacy"
[128] http://www.ncbop.org/Disciplinary%20Actions%20-%20PHARMACISTS/WeaverBradley 12708.pdf#search="compounding pharmacy"

[129] http://www.ncbop.org/Disciplinary%20Actions%20-%20PHARMACIES/DeepRiverDrug08944.pdf#search="compounding pharmacy"
[130] http://www.ncbop.org/Disciplinary%20Actions%20-%20PHARMACIES/RoyalPalmCompoundingPhcy10543.pdf#search="compounding pharmacy"
[131] http://www.ncbop.org/Disciplinary%20Actions%20-%20PHARMACIES/NewEnglandCompoundingCtr08439.pdf#search="compounding pharmacy"
[132] http://www.health.ri.gov/discipline/PHAApothecareCompoundingSolutions.pdf

In: Drug Compounding Pharmacies
Editor: Seth Sharpe
ISBN: 978-1-62808-176-3

Chapter 5

THE NEW ENGLAND COMPOUNDING CENTER AND THE MENINGITIS OUTBREAK OF 2012: A FAILURE TO ADDRESS RISK TO THE PUBLIC HEALTH*

Staff Report of the Senate Health, Education, Labor, and Pensions Committee

INTRODUCTION

On September 26, 2012, as a result of the rapid work of the Tennessee Department of Public Health and the Centers for Disease Control and Prevention (CDCP), an outbreak of an unusual strain of fungal meningitis was identified.

Preservative free methylprednisolone acetate (MPA), administered via spinal injection, was quickly identified as a likely source of the infections. The MPA was traced back to a compounding pharmacy in Framingham Massachusetts, the New England Compounding Pharmacy Inc., doing business as the New England Compounding Center (NECC). The Food and Drug Administration (FDA) subsequently determined that three separate lots of MPA, totaling over 17,000 doses produced by NECC between May 21, 2012

* This is an edited, reformatted and augmented version of the United States Senate Health, Education, Labor, And Pensions Committee publication, dated November 15, 2012.

and August 10, 2012, were contaminated with the *exserohilum rostratum* fungus.[i]

To date, NECC's failure to produce a sterile and safe product has led to more than 30 deaths and over 450 serious illnesses requiring treatment with high risk anti-fungal medications. The efforts of the CDCP and the Tennessee Department of Public Health allowed public health officials in 23 states to rapidly track and begin monitoring the approximately 14,000 possible recipients of the contaminated drug. But thousands of people around the country continue to wait and see whether they will develop meningitis, joint infections, spinal abscesses, or arachnoiditis. Those treated will face the risk of kidney and liver damage from the powerful anti-fungal drugs.

While the quick work of the public health community has led to early identification and treatment of many cases of meningitis, and reduced the fatalities resulting from the administration of the contaminated MPA, the Committee's investigation demonstrates that this crisis should have, and could have, been avoided entirely.

Since its creation in 1998, inspections of NECC by state, federal, and independent investigators have identified and documented profound deficiencies in the company's production of sterile drugs. The company has also been cited on multiple occasions for improper use of prescription blanks to solicit orders and failure to comply with state regulations requiring patient-specific prescriptions for compounded drugs.

Moreover, the same drug at issue in the current outbreak, NECC-produced MPA, had previously been a suspected cause of at least two cases with bacterial meningitis-like symptoms. These reports triggered an FDA inspection of the facility ten years prior to the current outbreak, in August 2002.

While the FDA sampling of NECC-produced MPA proved sterile at the time, other MPA samples were found to contain bacteria.[ii] As an FDA employee stated in a power point presentation to the Massachusetts Board of Registration in Pharmacy (Board) at the time, "Sterilization techniques and aseptic practices continue to raise questions, despite no positive (nonsterile) results from latest samples. Absence of evidence is not evidence of absence."[1]

Four years later in 2006, an independent evaluator reported to NECC manager and co-owner Barry Cadden that major areas of concern included "inadequate and incomplete documentation," that "end product testing is often

[i] Testing of the third lot is ongoing.

[ii] An outbreak of fungal meningitis caused by MPA compounded by a South Carolina pharmacy also occurred in mid-2002.

performed on 'stock solutions' and not the end product that is required," "process controls including validation of sterilization cycles and media fills are inadequate," and "in many cases the procedures are not in strict accordance with USP 795/797" as required by Massachusetts state law.[2]

In view of these repeated concerns with regard to the ability of NECC to safely produce compound drugs, it is difficult to understand why definitive action was never taken to either revoke its license or, at a minimum, closely monitor the company's operations. Instead, the company was allowed to grow and expand operations, ultimately holding licenses to ship drugs to at least 45 states. The same owners were subsequently permitted to open the far larger Ameridose, which supplied compounded drugs to hospitals around the country. Also, that company now has been found to lack adequate procedures to ensure that the compounds produced are safe, uniform or sterile.

This report is based on information obtained in the course of the Committee's investigation. It is intended to recount the known history of NECC, its related companies and their interactions with federal and state regulators as of November 15, 2012 to better understand the events leading to the current public health crisis.

THE NEW ENGLAND COMPOUNDING COMPANY

NECC was created in 1998 by the Conigliaro family. Three Conigliaro siblings and their spouses own the company: Douglas and Carla Conigliaro; Barry Cadden and Lisa Conigliaro Cadden; and Gregory Conigliaro. Ownership and management of the company have remained essentially unchanged since 1998. Pharmacists Barry Cadden and Lisa Conigliaro Cadden own 25 percent of the company, Carla Conigliaro owns 65 percent, and Gregory Conigliaro owns 10 percent. Gregory Conigliaro also owns a neighboring recycling business. Barry Cadden was in charge of operations and significant amounts of the actual compounding at NECC during the entire period of operations. The three siblings and spouses also own Ameridose and Alaunus, two companies created in 2006, in similar proportions.

NECC was granted a special pharmacy license by the Board in June 1998. That license allowed the company to produce compounded pharmaceutical products without operating a full-service pharmacy, but still subject to the state requirement that the company to have an individual patient prescription for each dose compounded. Massachusetts also adopted United States Pharmacopeia Standard <797>, which sets forth standards for compounding pharmacies including requirements for clean facilities, specific training for operators, and air quality evaluations.[3]

The first enforcement action against NECC began just 10 months after issuance of the license. In April of 1999, the Board filed a complaint against NECC for including blank prescriptions in solicitations to practitioners, a practice that violated state law. Six months later, in November of 1999, the Board resolved the complaint by issuing a warning to NECC in a private non-disciplinary advisory letter.[4]

In June 2001, the Idaho Board of Pharmacy complained to the Massachusetts Board that NECC was, among other things, including unapproved prescription forms in its solicitations to Idaho practitioners. Documents are unclear regarding whether the Board took formal action on this complaint. In fact, as detailed below, it appears that the Board has a dysfunctional system for logging incoming complaints and evaluating whether a complaint warrants assignment to an inspector.[5] Documents received by the Committee make clear, however, that NECC was investigated or warned for prescription-related concerns on at least 5 other occasions in the following 10 years.

Adverse Events

To the Committee's knowledge, the first time the safety of NECC's products was called into question was in early 2002. In March 2002, a prescribing doctor reported to the FDA that as many as five patients became ill following an epidural injection of NECC-produced betamethasone repositories.[6] He reported the illnesses to the FDA, alerted NECC about the issue, and returned unused doses to NECC without taking samples.[7] However, when the FDA arrived to inspect NECC on April 9, 2002, there were no records for the drugs in question.[8]

The FDA, joined for part of the inspection by the Board, spent three days inspecting NECC's facilities. When searching NECC's database, the FDA found a "date made" entry for the lot-number of drugs cited in the report but noted that "no associated records could be retrieved."[9] The FDA inspection report recounts that Barry Cadden asserted that the lot had never been produced but could provide no documentation that the lot had been cancelled.[10] Additionally, although the FDA contacted the physician making the report and confirmed he had returned the unused portion to NECC, FDA inspectors could find no record of the return.[11]

In the course of the inspection, FDA inspectors were told by Barry Cadden that approximately 4 lots of product produced between March and April 2002 had tested positive for endotoxin and were awaiting disposal.[12] FDA inspectors documented that NECC had sampled betamethasone repositories immediately after sterilization in the autoclave, and then left the product for up to 7 to 10 days before placing it in individual vials.[13] FDA inspectors reported an additional 8 areas of concern including a lack of procedures to ensure the operation of the autoclave, use of expired products, and inaccurate beyond use (i.e. expiration) dating.[14, iii]

In August of 2002, another series of adverse events were reported to the FDA.[15] These reports indicated that at least 2 patients were hospitalized for meningitis-like symptoms, and that the suspected sources of the infections were epidural injections of NECC-produced MPA, the same drug at issue in the current outbreak.[16]

The FDA, joined for part of the inspection by the Board, returned to NECC for a series of six days of inspections between October 2002 and February 2003. At that time Barry Cadden indicated to FDA inspectors that NECC was in the process of drastically expanding its operations. Since the FDA's prior inspection, NECC had doubled its square footage and hired additional staff. Further, NECC's manager stated his intent to expand sales to all 50 states, up from the 13 states in which it was then licensed.[17]

FDA tested unused vials of the MPA collected from the location of the adverse event report, and found that 5 of the 16 vials were contaminated with bacteria. The FDA also tested other vials obtained during inspections of NECC and found problems with super potent MPA and sub-potent betamethasone repositories.[18] Investigators again documented the use of procedures insufficient to ensure safe compounding. Those concerns included a "lack of documentation to verify that the autoclave itself is maintained and calibrated to perform its intended function," as well as a concern regarding a lack of safe procedures to ensure that "the transfer of bulk drug product and equipment from the autoclave... to another room è is not introducing contamination into the finished product."[19] The FDA's inspectors concluded, "Sample results revealed that the firm has sterility and potency issues with injectable steroid suspensions (betamethasone repository USP and methylprednisolone acetate USP)."[20]

[iii] Two days after the inspections, on April 18, 2002, the Nevada Board of Pharmacy submitted a complaint to the Board, alleging that NECC was selling non-FDA approved products in the state. It is unclear if the Board took any action as a result of this complaint.

In April 2002, prior to these inspections, the United States Supreme Court in *Thompson v. Western States Medical Center* ruled that section 503A of the Food Drug and Cosmetics Act included an impermissible restriction on commercial speech. The Supreme Court did not address provisions that clarified FDA's authority to regulate certain compound pharmacies, which the lower court held was not severable from the unconstitutional commercial speech restrictions. While NECC would likely have been subject to FDA regulation pursuant to section 503A of the Food Drug and Cosmetics Act, FDA's authority with regard to NECC under 503A was unclear after *Western States*, although FDA's general authority against unapproved new drugs, misbranded, or adulterated product was not in dispute. Despite the ambiguity regarding 503A, in May 2002 the FDA issued guidance which reasserted its authority to inspect compounding pharmacies and provided a non-exhaustive list of factors that the agency would consider in determining whether to take enforcement action when the scope and nature of a pharmacy's activities raise the kind of concerns ordinarily associated with drug manufacturing.

In this case, FDA took the position that the Board was better situated to take action against NECC. An FDA memo documenting a February 5, 2003, meeting between the FDA and the staff of the Board states that "a discussion was held to determine if NECC should be considered a manufacturer or a compounder" and that "current findings supported a compounding role."[21] The memo concludes:

> Mr. Elder [from FDA] concluded the meeting by summarizing the discussions and emphasizing the potential for serious public health consequences if NECC's compounding practices, in particular those relating to specific sterile products are not improved. The point was made that so long as a pharmacy's operations fall within the scope of the practice of pharmacyèFDA will generally defer to state authorities for regulatory oversight. In such cases FDA will seek to engage cooperative efforts aimed at achieving regulatory compliance and ensuring the safety and quality of compounded products.[22]

The FDA then officially stated in the NECC Inspection Report issued February 10, 2003, "[R]eferral to Massachusetts State Board of Pharmacy. Recommend firm be prohibited from manufacturing until they can demonstrate ability to make product reproducibly and dependably.

If state is unwilling to take action, recommend firm be enjoined for [Good Manufacturing Practices] deficiencies."[23]

Despite the formal recommendation that the state take action, it is unclear whether the Board took any additional action for the next year.[iv] It also does not appear that FDA conducted any follow-up to verify whether Massachusetts' response was sufficient to protect public health and safety.

Finally, on February 20, 2004, the Board staff conducted a compliance inspection and noted that NECC had taken corrective actions for the safety concerns identified in 2002 and 2003.[24] Nonetheless, the Board's staff recommended a public reprimand of NECC for its prior misconduct.[25]

On September 21, 2004, more than two years after the first reported cases of meningitis and other adverse events, and apparently acting on the staff recommendation, the Board voted to seek a public censure and probation for NECC's misconduct leading to the infections.[26,v] As was the Board's custom, they sent a consent decree to NECC that, if agreed to, would impose the relevant discipline and monitoring requirements for a three-year period.[27] The Board's staff transmitted its proposed consent decree to NECC on October 4, 2004.[28]

NECC did not agree to the proposed consent decree. NECC wrote to the Board asking it to instead consider non-public disciplinary action, to better protect NECC's business interest.[29] Counsel for NECC wrote: "once disclosed, the reprimand will surely result in inquiries/investigations in [other] jurisdictions. Regardless of the derivative actions taken, the attendant legal and administrative costs will be devastating."[30] The Board voted in November of 2004 to decline NECC's request for modifications to the consent decree.[31] Following that action by the Board, Committee interviews with Board staff suggest that the consent decree was referred for formal action to prosecuting attorneys within the Massachusetts Department of Public Health.[32]

For over a year, the record shows no formal order was filed and no hearing was held. Instead, it appears that attorneys for the Department of Public Health negotiated a modified consent agreement approved by the Board with an effective date of January 10, 2006.[33] The revised consent decree required that

[iv] FDA's investigation report also notes that Mr. Cadden, the manager of NECC, was serving on a committee for the state of Massachusetts, created to revise state regulations controlling compounding pharmacies. (FDA Inspectional Observations, Form FDA483, issued to Barry Cadden, R. Ph, Owner and Director of Pharmacy, New England Compounding Pharmacy, Inc., February 10, 2003.) The committee's work, however, became moot after the release of USP 797, which then was adopted by Massachusetts. 247 CMR 9.01(3).

[v] Between the Board staff report recommending censure and the Board vote to issue the consent decree, pharmacist Sophia Pasedis was appointed to the eleven-member Board. Ms. Pasedis appears to have been an employee of NECC in some capacity at the time of her appointment, and thus recused herself from the Board consideration of the consent decree. Ms. Pasedis is currently a manager of the Conigliaros' other drug company, Ameridose.

NECC submit to two inspections over a six-month period by a third-party evaluator, as well as a series of written assurances that recommended improvements had been made, in exchange for a suspended period of non-public probation.[34] The agreement was referred to as "non-disciplinary" and was "not reported to the National Association of State Boards of Pharmacy or other outside report agencies"[35]

Thus, almost four years after the two series of adverse events, including hospitalizations, likely caused by MPA, and three years after the FDA had stressed "the potential for serious public health consequences if NECC's compounding practices, in particular those relating to specific sterile products are not improved,"[36] the Board merely required NECC to hire an outside monitor, and made no mention of suspension or revocation of NECC's license.

PSI Monitoring

Pursuant to the revised consent decree, a third-party auditor, Pharmacy Support, Inc. (PSI) was selected to evaluate NECC's compliance with United States Pharmacopeia Standard 797, which the Board had recently adopted as the governing standard for Massachusetts.[37] PSI inspected NECC in January 2006 and noted multiple concerns, including sterility concerns. Among them, PSI noted a range of fundamental problems, including:

- NECC had "no requirements for donning proper attire or hand washing" when compounding medicines;
- "Mixing instructions are not specific and do not always indicate time and temperature;"
- "No quality control procedures are defined;"
- "Non-sterile 70% IPA is used to sanitize;"
- "Beard covers not worn;"
- "Hairnets and beard covers were not worn properly;"
- "Environmental monitoring procedures are inadequate;"
- "Calibrations are not performed properly;"
- "Floors in the unclassified/hybrid buffer area have not been sanitized in 3 months of use;"
- "[Beyond use dates] assigned incorrectly;"
- "There are no written procedures for receipt, storage, and accountability of controlled substances;"

- "[Standard Operating Procedures] are inadequate or not followed;"
- "Complaint forms were not available for some complaints logged in the complaint log;"
- "Most sections of complaint forms are not complete;"
- "Lot numbers are not assigned appropriately;"
- "4 out of 8 gloves observed had holes while CSP was compounded;" and
- Dry heat "sterilization equipment has not been verified."[38]

NECC took significant corrective measures, including replacing deficient equipment, conducting several training sessions for staff, and adopting a wide range of new standard operating procedures.[39] PSI submitted a final report on April 7, 2006, stating that NECC was largely compliant with pharmaceutical standards.[40,vi] In April and May, the Board received two more cursory letters from NECC assuring compliance with its remaining open issues.[41] On June 2, 2006, the Board informed NECC that it had fulfilled the requirements of its consent decree and that it considered the matter closed.[42]

ADDITIONAL NECC COMPLAINTS

At the time the Board acted to send the initial consent decree to NECC, it also acted to resolve three additional complaints against NECC in September 2004.[43] Despite ongoing investigations relating to serious adverse events, the Board issued three *non-disciplinary* private advisory letters to NECC resolving complaints submitted during the prior two years from practitioners in South Dakota, Texas, and Wisconsin.[44] While the advisory letters fail to spell out the specifics of the complaints, and the original complaints have not been reviewed by the Committee to date, it appears that NECC may have been soliciting bulk orders rather than patient-specific prescriptions, conduct that NECC was initially reprimanded for in 1999. A Board inspection report from

[vi] Six days later, at the end of an 8-week jury trial and three-year indictment, both PSI's CEO and Chief Compliance Officer were criminally convicted on 19 counts including fraud, mail fraud, and a violation of the Food and Drug Control Act. US v. Caputo, No. 03 CR 0126 (N.D. Ill. Oct 16, 2003). It is unclear how PSI was selected as the Massachusetts Board of Pharmacy has been unable to identify or produce documents discussing the selection of PSI in detail. Documents do show that PSI submitted a proposal to the Massachusetts Board and the prosecutor negotiating the consent agreement provided contact information to NECC's counsel.

around that time specifically notes that NECC "continues to reduce to writing orders on bulk purchase order forms and not on the approved prescription blanks. An issue previously addressed with Mr. Cadden."[45]

Additionally, in April 2004, five months before issuance of the advisory letters, the Board received a new complaint from a practitioner regarding the safety of NECC compounded triple anesthetic cream. The complaint states "My second concern is that [redacted] related to the purchasing technician that he would need a prescription for the product and that we could use the name of a staff member if we wanted to. He said 'other' institutions have used a nurses name....He assured her that it was legal. He indicated that after we received the product it was up to us how we used it and whom it was administered to."[46] It appears that this complaint triggered a Board inspection on November 2004. When questioned about the use of false names, Cadden responded "a review of the same documentation provided to you does show what would appear to be incorrect or repetitive names being provided by several of our prescribing physicians."[47] Yet the Board staff again recommended issuance of yet another non-disciplinary advisory letter dismissing this complaint.

On November 7, 2012, Department of Public Health officials informed the Committee that a July 2012 complaint against NECC, from the Colorado Board, for producing drugs in the absence of a patient-specific prescription had been discovered in the email of the Board's Executive Director. The complaint, which was received while the contaminated lots of MPA were still being produced by NECC, provided clear photographic evidence that NECC was shipping products in the absence of patient-specific prescriptions.[48] Further, Colorado had issued a cease and desist order to NECC in 2011 regarding this practice.[49] Board staff never acted on the July 2012 complaint, and it is unclear that the Board itself was aware of the complaint.[vii]

While Massachusetts state law requires that a compounding pharmacy possess a patient-specific prescription before preparing a compound drug, it appears that NECC has been consistently preparing and shipping batch products either in the absence of a prescription or to false prescription recipients since 1998. No regulatory entity appears to have undertaken a serious investigation of this ongoing practice, and the Board instead routinely dismissed and/or failed to act upon these repeated complaints.

[vii] The Board also informed the Committee that it had terminated the Executive Director and placed the Board's Counsel on administrative leave as a result of this discovery.

ADDITIONAL FDA ACTION

Two days after the Board finally voted to issue the consent decree in September 2004, the FDA and the Board returned to NECC, this time pursuant to a complaint regarding the company's improper production of an injectable dye used in ophthalmic procedures, Trypan Blue.[50] After Barry Cadden initially denied that any Trypan Blue was in stock, FDA inspectors located 189 vials of the product. Trypan Blue is commercially available and should not be compounded.[51]

This inspection led to the issuance of a December 4, 2006 FDA Warning Letter to NECC. The Warning Letter details issues including: the sale of compounded drugs without a patient-specific prescription; compounding copies of commercially-available drugs; selling misbranded compounding drugs; and compounding standardized non-approved drugs, with associated public health risks, on a large scale. It specifically notes that NECC "has reportedly told physicians' offices that using a staff member name on a prescription would suffice."[52]

While the FDA Warning Letter seeks corrective action within 15 days and threatens that failure to correct could result in further regulatory action including seizure or injunction, it does not appear that any further action was contemplated or that any efforts to ensure that corrective action were sought by the agency. Moreover, FDA chose to issue this Warning Letter without having learned from the Board what, if any, disciplinary actions had been taken in response to the inspections from October 2002-February 2003.[53] In January 2007 NECC responded to the Warning Letter, and in October 2008 the FDA re-asserted its authority to take "enforcement action, including seizure of the firm's products and/or an injunction against the firm and its principals" if violations noted in the Warning Letter were not corrected. The FDA also stated that "[i]n a future inspection, we will è verify that your firm's compounding practices are consistent with the policy articulated in the [Compliance Policy Guidelines.]"[54] This response came two years from the date of FDA's initial Warning Letter and four years from the date of the relevant inspection. FDA took no further action until the recent outbreak.

Further, a May 2011 email exchange shows that FDA staff, including the signatory to the October 2008 letter re-asserting FDA inspection authority, received a copy of a Colorado Cease and Desist Order issued to NECC in 2011 as the result of distribution of non-patient specific compounded drugs to hospitals in the Denver area. FDA staff apparently did not share the Cease and Desist order with the Massachusetts Board, or suggest that the Colorado Board do so until Colorado inspectors again discovered NECC stock compound drugs in another Colorado hospital in July 2012.[55]

INSPECTION FINDINGS SUBSEQUENT TO THE OUTBREAK

Unfortunately the long history of concerns was borne out in inspections by FDA and the Board following the 2012 fungal meningitis outbreak. The Massachusetts Board began a series of inspections of NECC on September 26, 2012. The Board and/or the FDA continued inspecting NECC from that date until October 26, 2012. The findings demonstrate a basic lack of compliance with USP <797> or with safe compounding as evidenced most clearly by the fact that "[v]isible black particulate matter was seen in several recalled sealed vials of Methylprednisolone Acetate."[56] Perhaps most critically, the FDA inspection found that NECC's environmental monitoring system documented 61 instances between January and August 2012 when either bacteria or mold was detected in concentrations exceeding action-level thresholds.[57]

The inspection reports found that while sterility testing conducted on the contaminated lots did not reveal unacceptably high levels of endotoxins, the sample provided was insufficient relative to the batch size. In fact when FDA sampled 50 vials of returned MPA it determined all fifty were contaminated with microbial growth despite the fact that sterility tests on one sample from the same lot in August 2012 has proven clear.[58] The FDA and Board inspections unsurprisingly again document a basic lack of procedure to ensure sterile products were being compounded safely including:

- Inspectors observed "greenish yellow discoloration" lining one of two autoclaves used to sterilize various components and equipment;[59]
- Inspectors observed "yellow residue lining the rear return of Weigh Station 2 Hood and greenish residue lining the rear return of Weigh Station 3 Hoodèused to weigh active ingredients and other raw materials;" [60]
- "Residual powder was visually observed within the [powder] hood during inspection;"
- "[Tacky] mats, which are used to trap dirt, dust, and other potential contaminants from shoes prior to clean room entryèwere visibly soiled with assorted debris;"[61] and

- "A leaking boiler adjacent to the requisite clean room created an environment susceptible to contaminant growth."[62]

The inspections also documented a continued disregard for the requirements of a patient-specific prescription for each compounded product. The state's preliminary investigation report noted: "NECC distributed large batches of compounded sterile products directly to facilities apparently for general use rather than requiring a prescription for an individual patient."

AMERIDOSE

One month after the terms of NECC's 2006 consent decree were deemed satisfied, the Board approved a license for a new company, Ameridose, owned by the Conigliaro family.[63], [viii] Massachusetts Board Member Sophia Pasedis has been a manager of record for the company.[64] According to media reports, Douglas Conigliaro, although not listed as an owner or manager, plays a significant role at Ameridose.[65]

Ameridose is also a sterile compounding company, but because it produces batch drugs for hospitals rather than patient-specific prescriptions, it is registered as a manufacturer with the FDA as well as with the Massachusetts Board.[66] The company does not manufacture any FDA approved product but rather is exclusively a large-scale compounder.

Until the outbreak, Ameridose contracted with Novation, the largest group purchasing organization in the country. Thus, Ameridose products were available to Novation's 3,000 hospital members as well as 22,000 other providers and facilities. Despite the history of problems with NECC and the joint ownership of the two companies, neither the licensure of Ameridose nor the large scale of its operations appears to have raised any concerns amongst the Board or the Board staff. Documents suggest that Ameridose was subject to routine pre-announced inspections by the Board in 2008 and 2011.[67]

However, the FDA had serious concerns with Ameridose. The FDA inspected the company in 2008 and found serious problems with the company's operations. Despite the large scale of Ameridose's operations even in 2008, investigators documented that products were shipped immediately

[viii] The ownership distribution is essentially the same with Carla Conigliaro owning 65 percent, Barry and Lisa Cadden owning 25 percent and Gregory Conigliaro owning 10 percent. (11-9-12 HELP Committee staff interview with NECC attorneys.)

without waiting for the results of sterility testing, that testing for potency and dose uniformity is not routinely performed and procedures were insufficient, and that the company was generally not in compliance with the requirements of USP 797 as required by Massachusetts law. As an example, management could not locate test results for 3 of 17 active ingredients inspected. Results of sampling tests taken at the August 2008 inspection returned a finding of superpotent Oxytocin, resulting in a recall of the product and an additional inspection in September 2008.[68] FDA staff placed Ameridose on the work plan for high risk facilities and recommended that a warning letter be issued to the company although no such letter was actually issued.[69]

While Ameridose was also the subject of at least 9 reports to the FDA of adverse events, faulty products, or medication errors, it is unclear that any of these triggered an inspection or investigation.[70] Following NECC's identification as the source of the fungal meningitis, the Board secured a temporary stop of Ameridose operations, though the company continues to hold a valid license. After the FDA began inspections on October 31, 2012, Ameridose issued a voluntary recall of all products.[71]

On November 12, 2012, the FDA issued a preliminary inspection report for Ameridose, finding startlingly similar problems to those they found NECC. Although the FDA has not reported any findings of contaminated drugs from Ameridose, the agency's preliminary findings "raised concerns about a lack of sterility assurance for products produced at and distributed by this facility."[72]

The FDA's inspection found that, like at NECC, there were clear problems with ensuring that drugs were sterile, or that doses were uniform. The FDA found that batches of drug product were not tested to ensure sterility, and that procedures were not established, written, or followed to prevent microbiological contamination of sterile drug products. What procedures were available did not include adequate validation of sterilization. The report also notes that the company failed to write or follow procedures detailing other aspects of their business.[73]

Moreover, the FDA found that testing of Ameridose's product did not include appropriate laboratory determination of conformance to the identity and strength of each active ingredient. And there were no written procedures for production and process controls to assure that the drug products had the identity, strength, quality and purity they purported to possess.[74]

Additionally, the FDA found that the buildings were not in good repair, that equipment and utensils were not cleaned, maintained, and sanitized at appropriate intervals to prevent contamination. The company further lacked suitable procedures to facilitate cleaning and maintenance, lacked equipment for adequate control over air pressure, and were infested with vermin.[75]

Conclusion

Given the history of NECC, the fact that the company produced and shipped a contaminated product that has led to 32 deaths and 461 infections to date is not a surprise. The surprise is that they were allowed to continue to engage in drug compounding for over a decade with this record.

Both federal and state regulators were well aware that NECC and its owners posed a risk to the public health. Both had documented that the company routinely flouted requirements that it compound products only when a patient specific prescription was received, compounded unapproved and commercially available products, potentially destroyed documents and samples relevant to adverse events, and most critically, repeatedly failed to demonstrate that the company could safely compound sterile products. There were a number of authorities and mechanisms for both federal and state regulators to address this issue, but bureaucratic inertia appears to be what allowed a bad actor to repeatedly risk public health.

The Committee will continue its investigation to determine how this tragic failure of oversight occurred, and how it can best be prevented in the future.

End Notes

[1] FDA Internal Memorandum, February 24, 2003, re: February 5, 2003 Meeting with Massachusetts Board of Pharmacy / Division of Professional Licensure (239 Causeway Street, Boston, MA 02114), p. 10 of Attachment 1.

[2] Letter from Pharmacy Support, Inc., to New England Compounding Center, January 30, 2006, and enclosure: Compounding Sterile and Non Sterile Preparations Observations and Recommendations Final Report.

[3] 247 CMR 9.01(3).

[4] Committee staff interview with Board inspectors 11/9/12.

[5] Senate HELP Committee staff interviews of Massachusetts Department of Public Health staff, October and November, 2012.

[6] FDA Investigative Report, re: FACTS 298826, New England Compounding Pharmacy Inc. EI 4/9, 4/10, 4/16/02; FDA Inspectional Observations, Form FDA483, issued to Barry Cadden, R. Ph, Owner and Director of Pharmacy, New England Compounding Pharmacy, Inc., April 16, 2002.

[7] FDA Investigative Report, re: FACTS 298826, New England Compounding Pharmacy Inc. EI 4/9, 4/10, 4/16/02.

[8] Id.

[9] FDA Inspectional Observations, Form FDA483, issued to Barry Cadden, R. Ph, Owner and Director of Pharmacy, New England Compounding Pharmacy, Inc., April 16, 2002, p. 4.

[10] Id. at 4.

[11] Id. at 4.

[12] Id. at 7.

[13] Id. at 7.

[14] Id. at 8.

[15] FDA Establishment Inspection Report and Continuation Sheet, re: FACTS 332851, New England Compounding Pharmacy Inc., EI Start: 10/24/02, EI End: 2/10/03; FDA Inspectional Observations, Form FDA483, issued to Barry Cadden, R. Ph, Owner and Director of Pharmacy, New England Compounding Pharmacy, Inc., February 10, 2003, p. 5.

[16] FDA Establishment Inspection Report and Continuation Sheet, re: FACTS 332851, New England Compounding Pharmacy Inc., EI Start: 10/24/02, EI End: 2/10/03, pp. 4-5.

[17] Id.

[18] Id.

[19] FDA Inspectional Observations, Form FDA483, issued to Barry Cadden, R. Ph, Owner and Director of Pharmacy, New England Compounding Pharmacy, Inc., February 10, 2003, p. 1.

[20] Id. at 2.

[21] FDA Internal Memorandum, February 24, 2003, re: February 5, 2003 Meeting with Massachusetts Board of Pharmacy / Division of Professional Licensure (239 Causeway Street, Boston, MA 02114), p. 2.

[22] Id. at 3-4.

[23] FDA Establishment Inspection Report, re: FACTS 332851, New England Compounding Pharmacy Inc., EI Start: 10/24/02, EI End: 2/10/03, p. 1.

[24] Massachusetts Department of Public Health – Division of Health Professions Licensure, Investigation Report, Re: New England Compounding Center, DS 03 055, and Barry Cadden, PH 03 066, March 4, 2004, p. 9.

[25] Id. at 9.

[26] Massachusetts Board of Registration in Pharmacy, Pharmacy Board Meeting Minutes: Tuesday, September 21, 2004, p. 9.

[27] Letter from the Massachusetts Board of Registration in Pharmacy to New England Compounding Center, re: Docket Number DS-03-055/PH-03-066/ New England Compounding Center (Permit #2848) and Barry Cadden, R.Ph., License No. 21239, October 4, 2004; Massachusetts Board of Registration in Pharmacy, Consent Agreement re: Docket No. DS-03-055, PH-03-066.

[28] Id.

[29] Letter from Paul Cirel, counsel for New England Compounding Center, to Board of Registration in Pharmacy, re: Docket Number DS-03-055/PH-03-066/ New England Compounding Center (Permit #2848) and Barry Cadden, R.Ph., License No. 21239, November 11, 2004.

[30] Id. at 4.

[31] Massachusetts Board of Registration in Pharmacy, Pharmacy Board Meeting Minutes: Tuesday, November 23, 2004, p2.

[32] Senate HELP Committee staff interviews of Massachusetts Department of Public Health staff, October and November, 2012.

[33] Massachusetts Board of Registration in Pharmacy, Consent Agreement re: Docket No. DS-03-055, PH-03-066, DS-05-040, January 10, 2006.

[34] Id.

[35] Id. at 1.

[36] FDA Internal Memorandum, February 24, 2003, re: February 5, 2003 Meeting with Massachusetts Board of Pharmacy / Division of Professional Licensure (239 Causeway Street, Boston, MA 02114), p. 2.

[37] Letter from Pharmacy Support, Inc., to New England Compounding Center, January 30, 2006, and enclosure: Compounding Sterile and Non Sterile Preparations Observations and Recommendations Final Report.

[38] Id.

[39] Pharmacy Support, Inc., Final Report USP <795>/<797> Implementation New England Compounding Center Framingham, Ma, April 7, 2006.

[40] Id.

[41] Letters between Board and NECC, April 12, 2006, April 19, 2006, May 10, 2006, and May 22, 2006.

[42] Letter from MA Board to NECC, June 2, 2006.

[43] Massachusetts Board of Registration in Pharmacy, Pharmacy Board Meeting Minutes: Tuesday, September 21, 2004.

[44] Letter from the Massachusetts Board of Registration in Pharmacy to New England Compounding Center, re: In the matter of DS-03-038 and PH-03-042 – New England Compounding Center (Permit #2848), September 30, 2004; Massachusetts Board of Registration in Pharmacy, Advisory Letter re: Docket No. DS-03-036, PH-03-042, September 30, 2004; Letter from the Massachusetts Board of Registration in Pharmacy to New England Compounding Center, re: In the matter of DS-03-060 and PH-03-070 – New England Compounding Center (Permit #2848), September 30, 2004; Massachusetts Board of Registration in Pharmacy, Advisory Letter re: Docket No. DS-03-060, PH-03-070, September 30, 2004; Massachusetts Board of Registration in Pharmacy, Advisory Letter re: Docket No. DS-04-062, PH-04- 161, September 30, 2004.

[45] Massachusetts Department of Public Health Å Division of Health Professions Licensure, Investigation Report, Re: New England Compounding Center, DS 03 055, and Barry Cadden, PH 03 066, March 4, 2004 at 7.

[46] Email from [redacted] to Massachusetts Board of Registration in Pharmacy, re: New England Compounding Center Activity in the State of Wisconsin, April 27, 2004.

[47] Massachusetts Department of Public Health Å Division of Health Professions Licensure, Investigation Report, Re: New England Compounding Center, DS 05 040, reviewed by Board Members on November 23, 2004.

[48] E-mail from Colorado Department of Regulatory Agencies to Massachusetts Board of Registration in Pharmacy, re: New England Compounding Center, July 26, 2012; Colorado State Board of Pharmacy, Special Report, re: New Egland Compounding Pharmacy, Inc. (WHO 7832), July 20, 2012.

[49] Id.

[50] FDA Internal Memorandum, Re: Inspection/Investigation of New England Compounding Center, January 26, 2005.

[51] Id. at 2.

[52] FDA Warning Letter to NECC, Dec 4, 2006.

[53] FDA Internal Memorandum, Re: Inspection/Investigation of New England Compounding Center, January 26, 2005 at 4.

[54] Letter from FDA to NECC, Oct 31, 2008.

[55] E-mails between Colorado Board of Pharmacy and FDA, May 10, 2011 and July 16, 2012; Colorado State Board of Pharmacy, Special Report, re: New England Compounding Pharmacy, Inc. (WHO 7832), July 20, 2012.

[56] Massachusetts Board of Registration in Pharmacy Report, NECC Preliminary Investigation Findings, October 23, 2012, at 4.

[57] FDA Inspectional Observations, Form FDA483, issued to Barry Cadden, Owner, New England Compounding Pharmacy, Inc., d/b/a/ New England Compounding Center, October 26, 2012.

[58] FDA Inspectional Observations, Form FDA483, issued to Barry Cadden, Owner, New England Compounding Pharmacy, Inc., d/b/a/ New England Compounding Center, October 26, 2012.

[59] Massachusetts Board of Registration in Pharmacy Report, NECC Preliminary Investigation Findings, October 23, 2012 at 1.

[60] Id. at 7.

[61] Massachusetts Board of Registration in Pharmacy Report, NECC Preliminary Investigation Findings, October 23, 2012 at 4.

[62] Id. at 5.

[63] Ameridose, LLC, Application for a New Store Å 50 Fountain Street, 2006.

[64] Id.

[65] Abby Goodnough et. al., "Spotlight Put on Founders of Drug Firm in Outbreak," New York Times, October 24, 2012, http://www.nytimes.com/2012/10/25/health/with-meningitis-outbreak-a-spotlight-onfamily-behind-compounding-pharmacy.html?pagewanted=all (accessed November 14, 2012).

[66] See, e.g. FDA Establishment Inspection Report, Ameridose, LLC, January 16, 2008, p. 1; Ameridose, LLC, Application for a New Store Å 50 Fountain Street, 2006.

[67] The Commonwealth of Massachusetts, Division of Health Professions Licensure, Inspection Report, 11/19/08; The Commonwealth of Massachusetts, Division of Health Professions Licensure, Inspection Report, 11/7/11.

[68] FDA Establishment Inspection Report, Ameridose, LLC, EI Start: 09/17/2008, EI End: 09/18/2008.

[69] FDA Establishment Inspection Report, Ameridose, LLC, August 22, 2008, p. 1.

[70] FDA, FAERS search results for suspect drugs labeled as Ameridose, New England Compounding Center or Alaunus, Reports initially received by FDA from 1/1/02 to 9/25/12, provided to Committee on November 8, 2012; See also, Letter from Wiley Rein, LLP to Massachusetts Board of Registration in Pharmacy, re: Complaint Against Ameridose LLC for Unlawful Manufacturing and Distribution of Pre-Mixed Nicardipine Injection Products, June 30, 2010 (the resulting investigation was administratively closed).

[71] FDA Press release: FDA reports voluntary recall of all Ameridose drug products, October 31, 2012, http://www.fda.gov/NewsEvents/Newsroom/PressAnnouncements/ucm326361.htm, (accessed November 14, 2012).

[72] Id.

[73] FDA Inspectional Observations, Form FDA483, issued to Gary Conigliaro, Vice President and General Manager, Ameridose, LLC, November 9, 2012.

[74] Id.

[75] Id.

In: Drug Compounding Pharmacies
Editor: Seth Sharpe

ISBN: 978-1-62808-176-3

Chapter 6

SELECTED RESOURCES ON FEDERAL OVERSIGHT OF COMPOUNDING PHARMACIES*

Judith M. Glassgold, Susan Thaul and Janet Kinzer

INTRODUCTION

In September 2012, the Centers for Disease Control and Prevention (CDC), the Tennessee Department of Health, the Food and Drug Administration (FDA), and other state health departments[1] began investigating a rare, non-contagious[2] outbreak of fungal meningitis. As of March 4, 2013, 20 states have reported 720 infections (including fungal meningitis and other conditions) and 48 deaths.[3] Patients at risk are those who received injections of contaminated, preservative-free methylprednisone acetate produced by the New England Compounding Center (NECC) after May 2012.[4] On October 4, 2012, the FDA verified that NECC was the source of the contaminated products.[5]

There has been extensive news coverage of the outbreak, which raised issues about the safety of compounded drugs and the role of federal and state

* This is an edited, reformatted and augmented version of a Congressional Research Service publication, CRS Report for Congress R42837, prepared for Members and Committees of Congress, from www.crs.gov, dated March 21, 2013.

governments in regulating compounded drugs and compounding pharmacies.[6] A number of policy questions about FDA authority and resources were raised in congressional hearings held by the Energy and Commerce Committee of the U.S. House of Representatives and the Health, Education, Labor, and Pensions Committee of the U.S. Senate.[7]

This report provides selected resources on the federal government's oversight of compounded drugs and compounding pharmacies, with an emphasis on relevant federal law and regulation and the activities of federal agencies related to and following the 2012 fungal meningitis outbreak. Resources from three states that have direct roles in this outbreak through licensing NECC, or reporting and tracking the fungal meningitis outbreak, are included as well. After a brief background section, the selected resources presented in this report appear in the following order:[8]

- federal law;
- congressional hearings;
- proposed legislation (current and previous Congresses);
- congressional support agency products;
- FDA agency guidance and regulations, meetings, and related documents;
- FDA actions on drug compounding;
- other federal agency actions and resources;
- selected state resources; and
- articles from legal, professional, and scientific journals.

BACKGROUND[9]

Drug compounding is a process in which a pharmacist combines, mixes, or alters various drug ingredients to create a medication for an individual patient in response to a practitioner's prescription.[10] It is generally used to prepare medications that are not typically commercially available, such as a drug in a lower dosage for a child, or a drug without a dye or a preservative in response to a patient allergy.[11] Compounding is a component of the practice of pharmacy and has historically been regulated at the state level, along with the licensure of pharmacy professionals and the regulation and licensure of pharmacies.[12]

The FDA is a regulatory agency within the Department of Health and Human Services (HHS). A key FDA responsibility, under the Federal Food, Drug, and Cosmetic Act (FFDCA), is the regulation of the safety and effectiveness of drugs sold in the United States.[13] Notwithstanding state regulation of pharmacies and the practice of pharmacy, the "FDA has determined that it should seriously consider enforcement action" over activities of a compounding pharmacy "when the scope and nature of a pharmacy's activities raise the kinds of concerns normally associated with a drug manufacturer and result in significant violations of the new drug, adulteration, or misbranding provisions of the Act [FFDCA]."[14] The recent FDA activities related to the fungal meningitis outbreak are not its first related to drug compounding. In the past the FDA has addressed the compounding of specific drugs (e.g., hormone replacement therapies for women, Makena, and colchicine[15]) and contamination of compounded drugs.[16] The next sections detail the existing federal role with regard to compounded drugs and compounding pharmacies.

FEDERAL LAW

Congress moved to clarify the FDA's role in the regulation of drug compounding by adding Section 503A (21 U.S.C. 353a) to the FFDCA as part of the FDA Modernization Act of 1997 (FDAMA, P.L. 105-115). Section 503A exempts compounded drugs from FFDCA requirements regarding drug adulteration, misbranding, and new drug approval process, provided that certain conditions are satisfied.[17]

Specifically, a drug product must be compounded by a licensed pharmacist or physician for an identified individual patient based on a valid prescription that was not solicited.[18] The compounded drug must comply with certain U.S. Pharmacopoeia standards or a relevant national formulary monograph, or be made from FDA-approved drug ingredients, and must meet certain manufacturing criteria.[19]

The drug compounded must not be one that appears on a list of drugs or drug products (published by the Secretary of HHS) that have been withdrawn or removed from the market because the product or components of the product have been found to be unsafe or not effective.[20]

Finally, the drug compounder may not "compound regularly or in inordinate amounts ... any drug products that are essentially copies of a commercially available drug product."[21]

Section 503A of FFDCA also states that a drug may be compounded and subject to the exemptions only if the pharmacy, licensed pharmacist, or licensed physician does not advertise or promote the compounding of any particular drug.[22] This was challenged by several compounding pharmacies as an impermissible regulation of speech under the First Amendment. The case went to the 5th and 9th Courts of Appeals and eventually to the U.S. Supreme Court, which agreed that the regulation was unconstitutional as it was not a permitted restriction of speech.[23] Whether the rest of the provisions of Section 503A are severable from the free speech provision, and are thus binding, is uncertain.[24]

- *FFDCA Section 503A* (21 U.S.C. 353a), Food and Drugs, Drugs and Devices: Pharmacy Compounding, http://www.gpo.gov/fdsys/pkg/ USCODE-2011-title21/ pdf/USCODE-2011-title21-chap9-subchapV-partA-sec353a.pdf.
- Specifics regarding Congressional intent during consideration of this issue can be found in Senate Committee on Labor and Human Resources report, S.Rept. 105- 43, *Food and Drug Administration Modernization and Accountability Act of 1997*, S. 830, 105th Cong., 1st sess., July 1, 1997, pp. 67-69, http://www.gpo.gov/ fdsys/pkg/CRPT-105srpt43/pdf/ CRPT-105srpt43.pdf.

Section 510 of FFDCA requires the registration of establishments involved in the manufacture, propagation, compounding, or processing of a drug or drugs; however, it specifically excludes pharmacies functioning in conformance with local laws.

- *FFDCA Section 510* (21 U.S.C. 360), Food and Drugs, Drugs and Devices: Registration of Producers of Drugs or Devices, http://www.gpo.gov/fdsys/pkg/ USCODE-2011-title21/pdf/USCODE-2011-title21-chap9-subchapV-partAsec360.pdf.

Congressional Hearings

The House Energy and Commerce Committee and the Senate Committee on Health, Education, Labor, and Pensions held hearings on the 2012 fungal meningitis outbreak and its relationship to FDA authority to regulate compounded drugs and compounding pharmacies. These hearings and related documents are listed below.

- November 15, 2012, "Pharmacy Compounding: Implications of the 2012 Meningitis Outbreak," hearing of the Senate Health, Education, Labor, and Pensions Committee, http://www.help.senate.gov/ hearings/hearing/?id=5f5def0d5056-a032-5297-eab57634d209.
  Materials available on the committee website include the hearing video, resources, and testimony.
  - United States Senate: Health, Education, Labor, and Pensions (HELP) Committee: Committee Staff Report, "The New England Compounding Center and the Meningitis Outbreak of 2012: A Failure to Address Risk to the Public Health," November, 15, 2012, http://www.help.senate.gov/ imo/media/ doc/11_15_12% 20HELP%20Staff%20Report%20on% 20Meningitis%20Outbrea k.pdf.
  - Resources and source materials compiled by the committee staff: http://www.help.senate.gov/imo/media/doc/Attachments.pdf.
  - Statement of Margaret Hamburg, Commissioner of Food and Drugs, at the Committee on Health, Education, Labor, and Pensions, U.S. Senate, November 15, 2012, http://www.help. senate.gov/imo/ media/ doc/ Hamburg3.pdf.
  - Other witness statements are available at http://www.help.senate. gov/hearings/hearing/?id=5f5def0d-5056-a032-5297-eab57634d209.
- November 14, 2012, "The Fungal Meningitis Outbreak: Could It Have Been Prevented?" hearing of the House Energy and Commerce Oversight and Investigations Subcommittee, http://energycommerce. house.gov/hearing/fungalmeningitis-outbreak-could-it-have-been-prevented.
  Materials available on the committee website include the complete hearing video, resources, opening statements, and testimony.
  - House Energy and Commerce Committee Majority Staff: Background Memorandum, November 14, 2012, http://energy commerce.house.gov/sites/republicans.energycommerce.house.go v/files/Hearings/OI/20121114/HMTG-112-HHRG-IF02-20121114 -SD001.pdf.
  - Statement of Margaret Hamburg, Commissioner of Food and Drugs, at the Subcommittee on Oversight and Investigations, House Energy

and Commerce Committee, November 14, 2012, http://www.fda.gov/NewsEvents/Testimony/ucm327664.htm.
- o The hearing transcript is available at http://www.cq.com/doc/congressionaltranscripts-4175157; http://www.cq.com/doc/ congressionaltranscripts-4175202 (panel 2); and http://www.cq.com/doc/congressionaltranscripts-4175216 (panel 3).

- In past years, U.S. Senate committees have held hearings on different compounded drugs and compounding pharmacies.
  - o April 19, 2007, "Bioidentical Hormones: Sound Science or Bad Medicine," S. Hrg. 110-129, Senate Special Committee on Aging, http://www.gpo.gov/fdsys/pkg/CHRG-110shrg37150/pdf/CHRG-110shrg37150.pdf.
    This hearing includes testimony by Jacques Rossouw, Chief of the Women's Health Initiative Branch, National Heart, Lung and Blood Institute, National Institutes of Health; Steven Galson, Director, Center for Drug Evaluation and Research, Food and Drug Administration; and Eileen Harrington, Deputy Director, Bureau of Consumer Protection, Federal Trade Commission.
  - o October 23, 2003, "Federal and State Role in Pharmacy Compounding and Reconstitution: Exploring the Right Mix to Protect Patients," S. Hrg. 108-378, Senate Committee on Health, Education, Labor, and Pensions, http://www.gpo.gov/fdsys/pkg/CHRG-108shrg90129/pdf/CHRG108shrg90129.pdf.
    This hearing included testimony by Janet Heinrich, Director of Health Care—Public Health Issues, GAO, and Steven Galson, Deputy Director, Center for Drug Evaluation and Research, FDA.

Proposed Federal Legislation

This section includes selected proposed legislation on the federal regulation of compounded drugs and compounding pharmacies. To date, no legislation has been proposed in the 113th Congress. This section does not include bills that would expand payment or financing of certain compounded drugs for specific diseases or conditions.[25]

112th Congress

- *H.R. 6584*, Verifying Authority and Legality in Drug Compounding Act of 2012, Representative Markey.
- *H.R. 6638,* S.A.F.E Compounded Drug Act of 2012, Representative DeLauro.

110th Congress

- H.Con.Res. 342, Expressing the sense of Congress that the FDA's new policy restricting women's access to medications containing estriol does not serve the public interest, Representative Ross.
- A bipartisan group of members of the Senate Health, Education, Labor, and Pensions Committee created a discussion draft in 2007, which is listed below.
  - Senate Discussion Draft: "Safe Compounding Drug Act of 2007."[26]

    This draft elicited comments from a variety of interested parties, some of which are presented below.
  - American Pharmacists Association (APhA), letter from Executive Vice President John A. Gans to Senators Kennedy, Burr, and Roberts, "RE: Discussion Draft of 'Safe Drug Compounding Act of 2007,'" March 7, 2007, http://www.aphafoundation.org/AM/Template.cfm?Section= Home&ContentID=16068&template=/CM/ContentDisplay.cfm.

    o Larry D. Sasich, Consultant, Public Citizen's Health Research Group, "Comments on a Draft of the Safe Drug Compounding Act of 2007," March 5, 2007, http://www.citizen.org/publications/publicationredirect.cfm?ID=7519.
  - Consumers Union, "Groups support the intent of the Safe Drug Compounding Act of 2007," letter to Senate Committee on Health, Education, Labor, and Pensions from signed group of organizations, April 9, 2007, http://www.consumersunion.org/news/groups-support-theintent-of-the-safe-drug-compounding-act-of-2007/.

105th Congress

- *H.R. 1060*, Pharmacy Compounding Act, Representative Burr, March 21, 1997, http://www.congress.gov/cgi-lis/bdquery/z?d105: H.R. 1060:/.
- *H.R. 1411*, Drug and Biological Products Modernization Act of 1997 (Introduced in House), Representative Burr.
- *H.R. 1411*, Prescription Drug User Fee Reauthorization and Drug Regulatory Modernization Act (Reported in House), Representative Burr.

104th Congress

- *H.R. 598*, Pharmacy Compounding Preservation Act of 1994, Representative Brewster, January 20, 1995.
- *H.R. 3199*, Drug and Biological Products Reform Act of 1996, Representative Burr.

OTHER CONGRESSIONAL DOCUMENTS

This section includes letters and reports from Members of Congress relevant to the 2012 fungal meningitis outbreak, compounded drugs, and compounding pharmacies.

- Senators Harkin and Enzi, U.S. Senate: Health, Education, Labor, and Pensions Committee, November 19, 2012, letter to state boards of pharmacy, http://www.help.senate.gov/newsroom/press/ release/?id= 3f3af08d-35ed-435cabe0-86a53c84fb78&groups=Chair.
  This letter was sent to all state pharmacy boards requesting information on their states' regulation of compounding pharmacies.
- Senators Blumenthal, Stabenow, Franken, and Feinstein, November 15, 2012, letter to Marilyn Tavenner, Acting Director, Centers for Medicare and Medicaid Services (CMS), http://op.bna.com/hl.nsf/ id/bdmr-927rru/$File/Blumenthalletter.pdf.

This letter requests information on the coordination of information between the FDA and CMS regarding compounding drug violations and Medicaid and Medicare payment policies.

- Representatives Markey, Waxman, Dingell, Pallone, DeGette, and Eshoo, November 15, 2012, letter to Gene Dodaro, U.S. Comptroller General, Government Accountability Office (GAO), http://markey.house.gov/document/ 2012/letter-gao-group-purchasing-organization. This letter requests that GAO investigate whether contracting practices by Group Purchasing Organizations and others in the prescription drug supply chain have had effects on drug pricing, drug shortages, and demand for compounded drugs.
- Report of the Office of Representative Markey, "Compounding Pharmacies, Compounding Risk," October 29, 2012, http://markey.house.gov/document/2012/compounding-pharmacies-compounding-risk-report.
  This report explores issues in the current regulation of compounding pharmacies.

CONGRESSIONAL SUPPORT AGENCY PRODUCTS

Congressional Research Service (CRS)

The following CRS materials address legal issues in drug compounding as well as federal regulation relevant to the current issues posed by the 2012 fungal meningitis outbreak.

- CRS Legal Sidebar, "Meningitis Outbreak Sheds Light on Regulation of Compounded Drugs," by Jennifer Staman, October 19, 2012, http://www.crs.gov/analysis/legalsidebar/pages/details.aspx?ProdId=276.
- CRS Report R40503, *FDA's Authority to Regulate Drug Compounding: A Legal Analysis*, by Jennifer Staman, October 17, 2012.

Government Accountability Office (GAO)

This testimony by Janet Heinrich, Director, Health Care—Public Health Issues, was delivered to the Committee on Health, Education, Labor, and Pensions of the U.S. Senate (October, 23, 2003).

- GAO, "Prescription Drugs: State and Federal Oversight of Drug Compounding by Pharmacies," GAO-04-195T, October 23, 2003, http://www.gao.gov/assets/ 120/110456.pdf.

FDA AGENCY GUIDANCE AND REGULATIONS, MEETINGS, AND RELATED DOCUMENTS

Guidance and Regulations

Compliance Policy Guidance on Pharmacy Compounding CPG 460.200

The following FDA guidance addresses drug compounding.[27] FDA reissued this CPG after the appellate court decisions. (See "Federal Law.")

- FDA, "CPG Sec. 460.200, Pharmacy Compounding," Compliance Policy Guidance for FDA Staff and Industry, May 2002, http://www.fda.gov/ICECI/ComplianceManuals/CompliancePolicyGuidanceManual/ucm074398.htm.

In this guidance, excerpted below, FDA identifies its authorities over certain types of activities of compounding pharmacies—those akin to manufacturers—while otherwise deferring to the regulatory prerogative of state pharmacy boards.[28]

> Generally, FDA will continue to defer to state authorities regarding less significant violations of the Act related to pharmacy compounding of human drugs. FDA anticipates that, in such cases, cooperative efforts between the states and the Agency will result in coordinated investigations, referrals, and follow-up actions by the states. However, when the scope and nature of a pharmacy's activities raise the kinds of concerns normally associated with a drug manufacturer and result in significant violations of the new drug, adulteration, or misbranding

provisions of the Act, FDA has determined that it should seriously consider enforcement action.

Drugs Not to Be Compounded

This regulation lists drugs withdrawn or removed from the market for reasons of safety or effectiveness and which may not be compounded under the exemptions of Section 503A(a) of FFDCA.

- 21 CFR 216.24, Food and Drug Administration, Pharmacy Compounding, April 1, 2011, http://www.gpo.gov/fdsys/pkg/CFR-2011-title21-vol4/pdf/CFR-2011- title21-vol4-part216.pdf.

FDA Meetings

On December 19, 2012, the FDA held two meetings regarding state and federal roles in pharmacy compounding. FDA first had an intergovernmental meeting with state officials and then hosted a public meeting titled "Framework for Pharmacy Compounding: State and Federal Roles." This public meeting included a summary of the results of the earlier intergovernmental meeting on the different roles of state and federal governments in regulating compounded drugs and compounding pharmacies. The meeting materials, archived webcast, and transcript are available on the FDA website:

- FDA, "Framework for Pharmacy Compounding: State and Federal Roles," www.fda.gov/NewsEvents/MeetingsConferences Workshops/ ucm330617.htm.

Other FDA Documents

The following FDA documents address compounding activities.

- FDA, "The Special Risks of Pharmacy Compounding," posted May 31, 2007, page last updated August 9, 2012, http://www.fda.gov/ downloads/ForConsumers/ ConsumerUpdates/ucm107839.pdf.

- FDA, "CPG Sec. 460.100, Hospital Pharmacies—Status as Drug Manufacturer," Compliance Policy Guide, issued October 1, 1980, page last updated January 20, 2010, http://www.fda.gov/ ICECI/ ComplianceManuals/CompliancePolicyGuidanceManual/ucm074397. htm.

The documents below discuss FDA requirements for prescription drug manufacturers, including those relating to adverse event reporting, inspections, sterility, and Current Good Manufacturing Practices (CGMPs). These materials provide background information on the regulation of manufacturing practices considered important in assuring the safety and effectiveness of prescription drugs.

- FDA, "MedWatch, The FDA Safety Information and Adverse Event Reporting Programs," http://www.fda.gov/Safety/MedWatch/ default.htm.
- FDA, "Drug Applications and Current Good Manufacturing Practice (CGMP) Regulations," page last updated September 7, 2012, http://www.fda.gov/Drugs/ DevelopmentApprovalProcess/Manufacturing/ucm090016.htm.
- FDA, "Questions and Answers on Current Good Manufacturing Practice (CGMP) for Drugs," page last updated December 21, 2011, http://www. fda.gov/ Drugs/DevelopmentApprovalProcess/Manufacturing/ucm124740.htm.
- FDA, "Facts about Current Good Manufacturing Practices (CGMPs)," page last updated June 25, 2009, http://www.fda.gov/Drugs/ DevelopmentApprovalProcess/Manufacturing/ucm169105.htm.
- FDA, "Dosage Form Drug Manufacturers cGMPs (10/93); Guide to Inspections of Dosage Form Drug Manufacturers—CGMPRS," Inspection Guides, page last updated April 30, 2009, http://www.fda.gov/ ICECI/Inspections/InspectionGuides/ucm074927.htm.
- FDA, "Sterile Drug Substance Manufacturers (7/94); Guide to Inspections of Sterile Drug Substance Manufacturers," page last updated April 30, 2009, http://www.fda.gov/ICECI/Inspections/ Inspection Guides/ucm074930.htm.
- FDA, "Guidance for Industry: Sterile Drug Products Produced by Aseptic Processing—Current Good Manufacturing Practice," Center for Drug Evaluation and Research (CDER), Center for Biologics

Evaluation and Research (CBER), Office of Regulatory Affairs (ORA),Pharmaceutical CGMPs, September 2004, http://www.fda.gov/downloads/Drugs/GuidanceComplianceRegulatoryInformation/Guidances/UCM070342.pdf.

FDA ACTIONS ON DRUG COMPOUNDING

Pharmacy Compounding

The FDA maintains a webpage with information on pharmacy compounding with a variety of information, including news and updates; information for consumers; information for health care providers; regulatory and policy information; FDA actions; and related topics.

- http://www.fda.gov/Drugs/GuidanceComplianceRegulatoryInformation/PharmacyCompounding/default.htm.

Fungal Meningitis Outbreak

The FDA maintains a website with updates on the 2012 fungal meningitis outbreak.

- FDA, "Multistate outbreak of fungal meningitis and other infections," http://www.fda.gov/Drugs/DrugSafety/FungalMeningitis/default.htm.

Selected FDA information bulletins and updates on the 2012 fungal meningitis outbreak are listed below in descending chronological order:

- FDA, "Questions and Answers on Fungal Meningitis Outbreak," page updated periodically, http://www.fda.gov/Drugs/DrugSafety/ ucm 322735.htm.
- FDA, New England District Office to NECC Re: Inspections on October 1-2, 4- 5, 9, 15, and 26, 2012, Form FDA 483 Inspectional Observations, Date Issued: October 26, 2012, http://insidehealthpolicy.com/iwpfile.html?file=oct2012%2Fhe10262012_necc.pdf.[29]

- FDA, "New England Compounding Center (NECC) Potentially Contaminated Medication: Fungal Meningitis Outbreak," MedWatch: The FDA Safety Information and Adverse Event Reporting Program, October 5, 2012, http://www.fda.gov/Safety/MedWatch/ Safety Information/ SafetyAlertsforHumanMedicalProducts/ucm322849.htm.
- "New England Compounding Center Issues Voluntary Nationwide Recall of All Products," Firm Press Release, October 6, 2012, posted at FDA, http://www.fda.gov/Safety/Recalls/ucm322901.htm.
- CDC, "CDC and FDA Joint Telebriefing on Investigation of Meningitis Outbreak," Press Briefing Transcript, October 4, 2012, http://www.cdc.gov/ media/releases/2012/t1004_meningitis_outbreak.html.

The following is a selection of FDA announcements and actions relevant to different compounded products and the marketing of prescription drugs without required FDA approval. These documents were selected based on their relevance to policy issues in the federal oversight of compounded drugs.

- FDA, "FDA Reports Voluntary Recall of All Ameridose Drug Product," November 12, 2012, http://www.fda.gov/Drugs/DrugSafety/ FungalMeningitis/ ucm326384.htm.
- FDA, "Pharmacy Compounding," page last updated October 10, 2012, http://www.fda.gov/Drugs/GuidanceComplianceRegulatoryInformatio n/ PharmacyCompounding/default.htm.
- FDA, "Unapproved Drugs: Drugs Marketed in the United States That Do Not Have Required FDA Approval," page last updated August 22, 2012, http://www.fda.gov/Drugs/Guidance ComplianceRegulatory Information/ EnforcementActivitiesbyFDA/SelectedEnforcementActionsonUnappr ovedDrugs/ default.htm.
- FDA, "Updated FDA Statement on Compounded Versions of hydroxyprogesterone caproate (the active ingredient in Makena),"[30] Press Announcements, June 15, 2012, http://www.fda.gov/ News Events/Newsroom/ PressAnnouncements/ucm308546.htm.
- FDA, "FDA Statement on Makena," Press Announcements, November 8, 2011, http://www.fda.gov/NewsEvents/ Newsroom/ PressAnnouncements/ ucm279098.htm.

- FDA, "Guidance for FDA Staff and Industry: Marketed Unapproved Drugs— Compliance Policy Guide; Sec. 440.100. Marketed New Drugs Without Approved NDAs or ANDAs," September, 19, 2011, http://www.fda.gov/ downloads/Drugs/GuidanceComplianceRegulatoryInformation/Guidances/ ucm070290.pdf.
- FDA, "FDA Alerts Health Care Professionals of Infection Risk from Repackaged Elastin Intravitreal Injections," page last updated August 30, 2011, http://www.fda.gov/Drugs/DrugSafety/ucm270296.htm.
- FDA, "FDA Statement on Makena," Press Announcements, March 30, 2011, http://www.fda.gov/NewsEvents/Newsroom/ Press Announcements/ucm249025.htm.
- FDA, "News for Pharmacy Compounders in Regards to Possible Melamine Contamination of Pharmaceutical Components," page last updated 09/23/2009, http://www.fda.gov/Drugs/Guidance Compliance RegulatoryInformation/ PharmacyCompounding/ucm183052.htm.
- FDA, "Questions and Answers about FDA's Enforcement Action against Unapproved Injectable Colchicine Products," page last updated May 15, 2009, http://www.fda.gov/Drugs/ Guidance ComplianceRegulatoryInformation/ EnforcementActivitiesbyFDA/SelectedEnforcementActionsonUnapprovedDrugs/ ucm119642.htm.
- FDA, "FDA Takes Action Against Compounded Menopause Hormone Therapy Drugs," FDA News Release, January 9, 2008, http://www.fda.gov/NewsEvents/ Newsroom/PressAnnouncements/2008/ucm116832.htm.
- FDA, "Warning Letter," to New England Compounding Center, December 4, 2006, http://www.fda.gov/ICECI/Enforcement Actions/ WarningLetters/2006/ ucm076196.htm.

Other Federal Agency Actions and Resources

Centers for Disease Control and Prevention (CDC)

CDC is involved in the 2012 fungal meningitis outbreak due to its responsibility to protect public health, including through disease surveillance. CDC has taken specific actions related to the 2012 fungal meningitis outbreak.

The following is a selection of CDC announcements and actions chosen based on their relevance to policy issues in the federal oversight of compounded drugs.

CDC, "Multistate Fungal Meningitis Outbreak Investigation," http://www.cdc.gov/hai/outbreaks/meningitis.html.

Selected CDC information bulletins and updates, and *Morbidity and Mortality Weekly Reports (MMWR)*, are listed below.

- CDC, "Current Case Count," http://www.cdc.gov/hai/outbreaks/ meningitismap.html.
- "Multistate Outbreak of Fungal Infection Associated with Injection of Methylprednisolone Acetate Solution from a Single Compounding Pharmacy— United States, 2012," MMWR, October 19, 2012, http://www.cdc.gov/mmwr/ preview/mmwrhtml/mm6141a4.htm.
- CDC, "CDC and FDA Joint Telebriefing on Investigation of Meningitis Outbreak," Press Briefing Transcript, October 4, 2012, http://www.cdc.gov/ media/releases/2012/t1004_meningitis_outbreak.html.
- CDC, "Meningitis and Stroke Associated with Potentially Contaminated Product," CDC Health Advisory, October 4, 2012, http://emergency.cdc.gov/ HAN/han00327.asp.
- "Notes from the Field: Multistate Outbreak of Postprocedural Fungal Endophthalmitis Associated with a Single Compounding Pharmacy— United States, March-April 2012," MMWR, vol. 61, no. 17 (May 4, 2012), pp. 30-31, http://www.cdc.gov/mmwr/preview/ mmwrhtml/ mm6117a5.htm.

Department of Justice (DOJ)

DOJ is involved in drug compounding issues through its responsibility to enforce the federal laws of the United States, including those regulating prescription drugs, compounded drugs, and compounding pharmacies. DOJ has pursued legal action in the past related to pharmacies that compound drugs; the following list represents cases found on the DOJ website.

- U. S. Department of Justice (DOJ), "Dallas Compounding Pharmacy Owner Pleads Guilty in Connection with Misbranded Drug

Shipment," press release, April 24, 2012, http://www.justice.gov/opa/pr/2012/April/12-civ-526.html. The press release stated that the pharmacist and the corporation pleaded guilty to misdemeanor violations of FFDCA by sending sub and super potent drugs across state lines. The press release also stated that three patients died as a result.

- DOJ, "U.S. Seeks Permanent Injunction Against Florida-Based Drug Compounding Lab," press release, April 9, 2010, http://www.justice.gov/opa/pr/ 2010/April/10-civ-438.html. The press release stated that a civil suit was filed alleging that the pharmacy had been introducing adulterated, misbranded, and unsafe drugs into interstate commerce.
- DOJ, Drug Enforcement Administration, "Docket No. 04-8, Wedgewood Village Pharmacy; Revocation of Registration," Federal Register Doc. E6-4771, April 3, 2006, http://www.deadiversion.usdoj.gov/fed_regs/actions/2006/fr0403.htm. This case involved the FDA's authority to inspect compounding pharmacies whose activities are more akin to those of manufacturers. This case was resolved by the Third Circuit Court of Appeals in 2005.

SELECTED STATE RESOURCES

Massachusetts

The Board of Registration in Pharmacy of the Massachusetts Executive Office of Health and Human Services (EOHHS) licenses pharmacy professionals and pharmacies. The EOHHS maintains a website specific to the 2012 fungal meningitis outbreak and the New England Compounding Center (NECC) product recall and investigation.

- Massachusetts EOHHS, "NECC Product Recall and Investigation," http://www.mass
- Massachusetts EOHHS, "Patrick-Murray Administration Appoints Chair of Special Commission on Compounding Pharmacies; Emergency Regulations Passed to Enhance Oversight of Industry," Press Release, November 1, 2012, http://www.mass.gov/eohhs/gov/newsroom/press-releases/dph/chair-of-specialcommission-on-compounding-pharmacies.html.

- Massachusetts EOHHS, "Pharmacy Board Draft Emergency Regulations," November 1, 2012, http://www.mass.gov/eohhs/docs/dph/quality/boards/necc/121101-pharmacy-board-draft-emergency-regs.pdf.
- Massachusetts EOHHS, "New England Compounding Center (NECC): Preliminary Investigation Findings," Board of Registration in Pharmacy Report, October 23, 2012, http://www.mass.gov/eohhs/docs/dph/quality/boards/necc/necc-preliminary-report-10-23-2012.pdf.
- Massachusetts EOHHS, "Statement of Dr. Madeleine Biondolillo," Director of the [Massachusetts] Bureau for Health Care Safety and Quality, Joint DPH / FDA / CDC Teleconference Call, October 11, 2012, http://www.massdph/quality/boards/ pharma cy-121011-statement-madeleine-biondolillo.pdf.
- Massachusetts EOHHS, "Alert—New England Compounding Center Product Recall Information," Press Release, October 4, 2012, http://www.mass.gov/ eohhs/docs/dph/quality/boards/pharmacy-alert-necc.pdf.

COLORADO

Colorado took actions against NECC in 2011 and 2012 for the unlawful distribution of prescription drugs. The documents below are related to those actions.

- Colorado State Board of Pharmacy, "Special Report: New England Compounding Pharmacy, Inc. (WHO 7832)," July 20, 2012, http://www.mass.gov/eohhs/docs/dph/quality/boards/pharmacy/120720-coloradostate-board-of-pharmacy-report.pdf.
- E-mail correspondence between Colorado State Board of Pharmacy and Massachusetts Board of Registration of Pharmacy regarding NECC, July 2012, http://www.mass.gov/eohhs/docs/dph/quality/boards/pharmacy/120727-coloradostate-board-of-pharmacy-email.pdf.

Tennessee

Tennessee was one of the first states to identify the 2012 fungal meningitis outbreak. Its Department of Health maintains a website focusing on the 2012 fungal meningitis outbreak.

- Tennessee Department of Health, "Multistate Meningitis Outbreak Investigation," http://health.state.tn.us/outbreakinvestigation.shtml.

ARTICLES FROM LEGAL, PROFESSIONAL, AND SCIENTIFIC JOURNALS[31]

This section lists selected articles selected from a search of PubMed/Medline. The search terms focus on public health concerns and included variations of pharmacy compounding, compounded drugs, adverse effects, and fatalities.

Selected 2013 Articles on Compounded Drugs

- Beth. P. Bell and Rima F. Khabbaz, "Responding to the Outbreak of Invasive Fungal Infections: The Value of Public Health to Americans," *JAMA, The Journal of the American Medical Association*, vol. 309, no. 9, March 6, 2013, pp. 883-884, http://jama.jamanetwork.com/article.aspx?articleid=1567243.
- Joseph V. Pergolizzi Jr, Sumedha Labhsetwar, and Jo AnnLeQuang, "Compounding Pharmacies: Who is in Charge?," *Pain Practice*, vol. 13, no. 3, pp. 253-257, March, 2013; January 28, 2013, on-line, http://onlinelibrary.wiley.com/doi/10.1111/papr.12033/abstract.
- Roy Guharoy et al., "Compounding Pharmacy Conundrum: 'We Cannot Live Without Them But We Cannot Live With Them' According to the Present Paradigm," CHEST, February 14, 2013, http://journal.publications.chestnet.org/article.aspx?articleID=1570173 (abstract).

Selected Articles on the 2012 Fungal Meningitis Outbreak and Compounding Pharmacies

- Jeffrey M. Drazen, Gregory D. Curfman, Lindsey R. Baden, and Stephen Morrissey, "Editorial: Compounding Errors," New England

Journal of Medicine, November 8, 2012, http://www.nejm.org/doi/full/10.1056/NEJMe1213569.

- Kevin Outterson, "Perspective: Regulating Compounding Pharmacies after NECC," New England Journal of Medicine, November 7, 2012, http://www.nejm.org/doi/full/10.1056/NEJMp1212667.
- Marion A. Kainer et al., "Fungal Infections Associated with Contaminated Methylprednisolone in Tennessee," New England Journal of Medicine, November 6, 2012, http://www.nejm.org/doi/pdf/10.1056/NEJMoa1212972.
- Lucy E. Wilson et al., "Fungal Meningitis from Injection of Contaminated Steroids," New England Journal of Medicine, October 26, 2012, http://jama.jamanetwork.com/article. aspx?articleid=1387597.
- W.J. Zohler, "Steps to Improve Quality in the Compounding Lab," International Journal of Pharmaceutical Compounding, vol. 16, no. 2, March-April, 2012, pp. 136-144, http://www.ncbi.nlm.nih.gov/pubmed/23050326 (abstract).

Selected Articles Regarding Safety Issues and Compounding Pharmacies 2006 on

- S. Huang, et al., Notes from the field: Multistate outbreak of post procedural fungal endophthalmitis associated with a single compounding pharmacy, Morbidity and Mortality Weekly Report (MMWR), vol. 61, no. 17, March-April 2012, pp. 310-311, http://www.medscape.com/viewarticle/763654.
- Robert A. Goldberg et al., "An Outbreak of Streptococcus Endophthalmitis After Intravitreal Injection of Bevacizumab," *American Journal of Ophthalmology*, February 2012, pp. 204-208.
- Cheryl A. Thompson, "Bacteremia Outbreak Tied to Improper Filtration by Compounding Pharmacy," *American Journal of Health-System Pharmacy*, vol. 68, no. 21, November 15, 2011, pp. 2110-12.
- Neil Gupta et al., "Investigation of an outbreak of *Serratia marcescens*
- Bloodstream infections in patients receiving total parenteral nutrition—Alabama, 2011." Authors affiliated with CDC and Alabama Dept. of Health, Paper presented at conference IDSA, October, 2011, https://idsa.confex.com/idsa/2011/ webprogram/Paper31248.html.

- Helen M. Steel et al., "Fatal Respiratory Events Caused by Zanamivir Nebulization," *Clinical Infectious Diseases*, vol. 51, no. 1, July 1, 2010, p. 121, http://cid.oxfordjournals.org/content/ 51/1/121.1.full. pdf+html.
- Lisa A. Maragakis et al., "*Sphingomonas paucimobilis* Bloodstream infections associated with contaminated intravenous fentanyl," *Emerging Infectious Disease*, vol. 15, no. 1, 2009, pp. 12-18.
- Tamira Mullarkey, "Pharmacy compounding of high-risk level products and patient safety," *American Journal of Health-System Pharmacy*, vol. 66, no. 17, supp. 5, September 1, 2009, pp. S4-S13 [PDF pp. 5-14], http://www.ashp.org/ DocLibrary/CE/09004AJHP _suppl5_2009.pdf.
- Larry D. Sasich, "Unknown Risks of Pharmacy—Compounded Drugs," *Journal of the American Osteopathic Association*, vol. 108, no. 2, February 1, 2008, pp. 51-52.
- Bruce Patsner, "Pharmacy Compounding of Bioidentical Hormone Replacement Therapy (BHRT): A Proposed New Approach to Justify FDA Regulation of These Prescription Drugs," *Food and Drug Law Journal*, 2008 (63 Food Drug L.J. 459).
- Daniel Meron, "Legal Developments Relevant to FDA Authority," *Food and Drug Law Journal*, 2007 (67 Food Drug L.J. 441).
- Karen S. McCoy, "Compounded Colistimethate as Possible Cause of Fatal Acute Respiratory Distress," New England Journal of Medicine, vol. 357, November 29, 2007, pp. 2310-2311, http://www.nejm.org /doi/full/10.1056/NEJMc071717.
- N. J. McKeown et al., Deaths from intravenous colchicine resulting from a compounding pharmacy error—Oregon and Washington, 2007, October 12, 2007, 56(40) 1050-1052, http://www.cdc.gov/ mmwr/preview/mmwrhtml/ mm5640a3.htm.
- Rebecca H. Sunenshine et al., "A Multi-state Outbreak of Serratia marcescens Bloodstream Infection Associated with Contaminated Intravenous Magnesium Sulfate from a Compounding Pharmacy," Clinical Infectious Diseases, vol. 45, no. 5, September 1, 2007, pp. 527-533, http://cid.oxfordjournals.org/content/45/ 5/527.full.pdf+ html.
- Rachel Civen et al., "Outbreak of Serratia marecescens Infections Following Injection of Betamethasone Compounded at a Community Pharmacy," Clinical Infectious Diseases, vol. 43, no. 7, August 22,

2006, pp. 831-837, http://cid.oxfordjournals.org/content/43/7/831.full.pdf+html.

- Melissa R. Held et al., “Life-Threatening Sepsis Caused by Burkholderia cepacia From Contaminated Intravenous Flush Solutions Prepared by a Compounding Pharmacy in Another State,” Pediatrics, June 19, 2006, http://pediatrics.aappublications.org/content/118/1/e212.full.pdf+html.

End Notes

[1] Beth Bell, Director, Director, National Center for Emerging and Zoonotic Diseases, Centers for Disease Control and Prevention (CDC), Department of Health and Human Services (HHS), “Testimony Before the Health, Education, Labor, and Pensions Committee, U.S. Senate: The CDC and Public Health Response to the 2012 Fungal Meningitis and Other Infections Outbreak,” pp. 2-6, http://www.help.senate.gov/imo/media/ doc/Bell.pdf.

[2] This form of fungal meningitis is non-contagious; the more common form of meningitis is bacterial and is infectious. See http://www.cdc.gov/ meningitis/fungal.html.

[3] Centers for Disease Control and Prevention, Multistate Fungal Meningitis Outbreak—Current Case Count, http://www.cdc.gov/hai/outbreaks/ meningitis.html.

[4] Centers for Disease Control and Prevention, Multistate Fungal Meningitis Outbreak, http://www.cdc.gov/HAI/ outbreaks/currentsituation/.

[5] Beth Bell, Director, National Center for Emerging and Zoonotic Diseases, Centers for Disease Control and Prevention, HHS, “Testimony Before the Health, Education, Labor, and Pensions Committee, U.S. Senate: The CDC and Public Health Response to the 2012 Fungal Meningitis and Other Infections Outbreak,” p. 6, http://www.help.senate.gov/ imo/media/doc/Bell.pdf.

[6] Lena H. Sun and Kimberly Kindy, “FDA inspects specialty compounding pharmacies in targeted action,” Washington Post, March 1, 2012, http://articles.washingtonpost.com/2013-03-01/national/37368259_ 1_pharmacies-drugevaluation-inspection; Kimberly Kindy, Lena H. Sun, and Alice Crites, “Compounding pharmacies have been linked to deaths, illnesses and safety failures for years,” Washington Post, February 7, 2013, http://articles.washingtonpost.com/ 2013-02-07/national/36970682_ 1_medications-for-individual-patients-massachusetts-pharmacy-meningitis-outbreak; Andrew Pollack and Sabrina Tavernise, “Oversight Failures Documented in Meningitis Outbreak,” New York Times, November 21, 2012, http://www.nytimes.com/2012/11/22/ health/ documents-show-fdas-failures-in-meningitisoutbreak.html; Denise Grady, “Deaths Stir a Dispute on Powers of FDA,” New York Times, November 20, 2012, http://www.nytimes.com/2012/ 11/20/health/tainted-drug-deaths-spawn-heated-debate-over-fdas-powers.html?ref= health&_r=0; Sabrina Tavernise, “FDA Chief Seeks Expanded Authority to Improve Safety of Drug Compounders,” New York Times, November 14, 2012, http://www.nytimes. com/2012/11/15/health/fda-asking-for-more-control-overdrug- compounding.html?ref=health; and Eric Lichtblau and Sabrina Tavernise, “Friends in Congress Have Helped Drug Compounders Avoid Tighter Rules,” New York Times,

November 13, 2012, http://www. nytimes.com/2012/11/14/health/niche-drugmakers-get-help-on-capitol-hill.html?pagewanted=all&_r=0.

[7] U.S. Congress, House Committee on Energy and Commerce, Subcommittee on Oversight and Investigations, The Fungal Meningitis Outbreak: Could It Have Been Prevented, 112th Cong., 2nd sess., November 14, 2012, http://search.proquest.com/docview/1151991974/13AA2A2331776953950/1?accountid=41647. U.S. Congress, Senate Committee on Health, Education, Labor, and Pensions, Pharmacy Compounding: Implications of the 2012 Meningitis Outbreak, 112th Cong., 2nd sess., November 15, 2012, http://www.help.senate.gov/hearings/hearing/?id=5f5def0d5056-a032-5297-eab57634d209.

[8] Articles or sources listed in this report are available from the Congressional Research Service. Please contact Janet Kinzer, 7-7561.

[9] This section relies on CRS Report R40503, FDA's Authority to Regulate Drug Compounding: A Legal Analysis, by Jennifer Staman.

[10] Food and Drug Administration, "Pharmacy Compounding," page last updated March 1, 2013, http://www.fda.gov/ Drugs/GuidanceCompliance RegulatoryInformation/Pharmacy Compounding/default.htm. ; and see also Western States Medical Center v. Shalala, 238 F.3d 1090, 1092 (9th Cir. 2001).

[11] Western States Medical Center et al. v. Shalala, 69 F. Supp. 2d 1288, 1292 (D. Nev. 1999). See also Statement of Steven K. Galson, Acting Director, Center for Drug Evaluation and Research, U.S. Food and Drug Administration, before the Senate Committee on Health, Education, Labor, and Pensions, hearing on "Federal and State Role in Pharmacy Compounding and Reconstitution: Exploring the Right Mix to Protect Patients," (Oct. 23, 2003).

[12] Comments of Margaret Hamburg, Commissioner of Food and Drugs, FDA, Transcript of Public Meeting "Framework for Pharmacy Compounding: State and Federal Roles," December 19, 2012; and Western States Medical Center et al. v. Shalala, 69 F. Supp. 2d 1288, 1292 (D. Nev. 1999).

[13] For more information about the FDA's responsibilities with regard to prescription drugs, see CRS Report R41983, How FDA Approves Drugs and Regulates Their Safety and Effectiveness, by Susan Thaul.

[14] See section identified as "Policy" in Food and Drug Administration, Pharmacy Compounding: Compliance Policy Guidance, CPG Sec. 460.200, Washington, DC, May 2002, http://www.fda.gov/ICECI/ ComplianceManuals/CompliancePolicyGuidanceManual/ucm074398.htm.

[15] Makena is a product that is used to reduce the risk of preterm birth and colchicine is a drug used to treat gout.

[16] For more information, see "FDA Actions on Drug Compounding."

[17] See Section 127 of P.L. 105-115 (FFDCA 503A).

[18] 21 U.S.C. §353(a)

[19] 21 U.S.C. §353a(b)(1).

[20] 21 U.S.C. §353a(b)(1).

[21] 21 U.S.C. §353a(b)(1)(D).

[22] 21 U.S.C. §353a(c).

[23] Western States, 535 U.S.357 (2002).

[24] For more information on legal issues, see CRS Report R40503, FDA's Authority to Regulate Drug Compounding: A Legal Analysis, by Jennifer Staman.

[25] For example, see H.R. 232 in the 113th Congress; H.R. 6326, S. 3496 in the 112th Congress; and H.R. 1812 in the 110th Congress.

[26] Copies of this draft legislation are available from Janet Kinzer, CRS, 7-7561.

[27] Please note, agency rules have the force and effect of law; guidance documents are considered to be a general statement of policy.

[28] For a discussion of the legal and regulatory issues in light of legal challenges, see S. A. Coffina, "Coming Soon: Increased Scrutiny of Compounding Pharmacies," Bloomberg/BNA Pharmaceutical Law & Industry Report (November 16, 2012), and CRS Report R40503, FDA's Authority to Regulate Drug Compounding: A Legal Analysis, by Jennifer Staman.

[29] For the full text of articles, please contact Janet Kinzer, CRS, 7-7561.

[30] This prescription drug is used to reduce the risk of preterm birth.

[31] For the full text of articles, please contact Janet Kinzer, CRS, 7-7561.

Index

D

E

F

G

K

L

M

N

O